Strategic Studies Institute
and
U.S. Army War College Press

TERRORIST AND INSURGENT UNMANNED AERIAL VEHICLES: USE, POTENTIALS, AND MILITARY IMPLICATIONS

Robert J. Bunker

August 2015

*****

Comments pertaining to this report are invited and should be forwarded to: Director, Strategic Studies Institute and U.S. Army War College Press, U.S. Army War College, 47 Ashburn Drive, Carlisle, PA 17013-5010.

*****

All Strategic Studies Institute (SSI) and U.S. Army War College (USAWC) Press publications may be downloaded free of charge from the SSI website. Hard copies of this report may also be obtained free of charge while supplies last by placing an order on the SSI website. SSI publications may be quoted or reprinted in part or in full with permission and appropriate credit given to the U.S. Army Strategic Studies Institute and U.S. Army War College Press, U.S. Army War College, Carlisle, PA. Contact SSI by visiting our website at the following address: *www.StrategicStudiesInstitute.army.mil.*

*****

The Strategic Studies Institute and U.S. Army War College Press publishes a monthly email newsletter to update the national security community on the research of our analysts, recent and forthcoming publications, and upcoming conferences sponsored by the Institute. Each newsletter also provides a strategic commentary by one of our research analysts. If you are interested in receiving this newsletter, please subscribe on the SSI website at *www.StrategicStudiesInstitute.army.mil/newsletter.*

ISBN 1-58487-698-0

FOREWORD

This manuscript focuses on the present threat posed by terrorist and insurgent use of unmanned aerial vehicles (UAVs) as well as the associated future threat potentials identified. The work presents a counterintuitive analysis in the sense that armed drones are typically viewed as a component of America's conventional warfighting prowess—not a technology that would be used against U.S. troops deployed overseas or against civilians back home. Utilizing a red teaming approach, the author, Dr. Robert J. Bunker, a past Minerva Chair at our institution, investigates the emerging threat of such UAV use. His unique analysis and creative approach, especially when related to the threat scenario variants generated, make for very informative reading.

The work is divided into an introduction to the topical area, a UAV historical overview and discussion of present use by the U.S. military, a chronological narrative of terrorist and insurgent UAV use (and attempted use) from 1994 through 2015, the ensuing baselines and trending identified, and the foreseen potentials derived from these trends—based upon tactical, operational, and strategic influencing scenarios, and the resulting military implications and suggested policy responses this will entail. The analysis not only has immediate value for Army force protection and counterterrorism programs, but also for research being conducted on projected robot-on-human force-on-force engagements in insurgency type environments, as well as strategic considerations related to emerging "drone swarm" concepts and the changing character of warfare as robot Landpower technologies evolve and are increasingly fielded.

The Strategic Studies Institute hopes that the analysis and recommendations found in this monograph will be of use to the various U.S. Army organizations impacted by nonstate threat UAV use and those entities in sister services also so effected, as well as domestic policing and federal law enforcement bodies tasked with counterterrorism and homeland security missions. Further, other Army and sister service entities, as well as various U.S policymaking bodies, hopefully will find the larger implications posed by this report related to semi-autonomous and autonomous UAV type robotic systems of some benefit.

DOUGLAS C. LOVELACE, JR.
Director
Strategic Studies Institute and
U.S. Army War College Press

ABOUT THE AUTHOR

ROBERT J. BUNKER is 2015 Futurist in Residence, Behavioral Research and Instruction Unit, Federal Bureau of Investigation Academy, Quantico, VA, and an Adjunct Research Professor, Strategic Studies Institute, U.S. Army War College. He is also Adjunct Faculty, Division of Politics and Economics, Claremont Graduate University, and a Non-Resident Counterterrorism Fellow, TRENDS Research & Advisory, Abu Dhabi, United Arab Emirates. Past professional associations include Distinguished Visiting Professor and Minerva Chair at the Strategic Studies Institute, U.S. Army War College; Chief Executive Officer, Counter-OPFOR Corporation; Adjunct Faculty, School of Policy, Planning, and Development, University of Southern California; Terrorism Instructor, California Specialized Training Institute, California Office of Emergency Services; Staff Member (Consultant), Counter-OPFOR Program, National Law Enforcement and Corrections Technology Center-West; Fellow, Institute of Land Warfare, Association of the U.S. Army; Adjunct Faculty, National Security Studies M.A. Program and Political Science Department, California State University San Bernardino; and Faculty, Unconventional Warfare M.A. Distance Education Program, American Military University. Dr. Bunker has delivered over 200 presentations—including papers and training—to military, law enforcement, academic, and policy audiences including U.S. Congressional Testimony. He has hundreds of publications ranging from edited books and booklets through reports, chapters, articles and essays, response guidance, subject bibliographies, and encyclopedia entries in academic, policy, military and law enforcement venues. Among these are *Studies in Gangs and Cartels*, with John P. Sullivan (Routledge, 2013), and *Red Teams and Counter-terrorism Training* with Stephen Sloan, (University of Oklahoma, 2011); and edited (and co-edited) works including *Global Criminal and Sovereign Free Economies and the Demise of the Western Democracies* (Routledge, 2014), *Criminal Insurgencies in Mexico and the Americas: The Gangs and Cartels Wage War* (Routledge, 2012), *Narcos Over the Border: Gangs, Cartels and Mercenaries* (Routledge, 2011), *Criminal-States and Criminal-Soldiers* (Routledge, 2008), *Networks, Terrorism and Global Insurgency* (Routledge, 2005), and *Non-State Threats and Future Wars* (Routledge, 2002). Dr. Bunker holds university degrees in political science, government, social science, anthropology-geography, behavioral science, and history.

SUMMARY

Derived from the unmanned aerial vehicle (UAV) use threat scenarios, three levels of military significance are foreseen with terrorist and insurgent activities associated with these devices. Because of the technologies that will eventually be associated with UAVs—robotics and expert (and artificial intelligence) systems networked together—their significance is projected as increasing over time from the tactical to the operational and then to the strategic levels of concern. These levels of military implication and the suggested policy responses related to them are:

Tactical.

The impact of even singular terrorist UAV use at this level is viewed as an immediate- and near-term problem. It may represent more of a domestic security issue than an overseas basing or deployment threat—although such weaponized devices could just as easily be utilized for terrorism purposes overseas against service personnel and their families as they could be used against civilians in the United States. The tactical level threat derived from the drone-up shooting, improvised explosive device (IED) crowd targeting, and aircraft takedown scenario variants will be of concern to domestic law enforcement, homeland security, and Federal Bureau of Investigation Joint Terrorism Task Force elements as well as the military.

Operational.

This level of impact is insurgency environment focused and pertains to the use of groups of human controlled and semi-autonomous UAVs. The virtual martyrs and drone squadron scenario variants portrayed the various types of flying IED, weapons platform, and human insurgent fighter combined arms' hybrid threats that could be encountered at this level of concern. While conceivably representing a present-day operational risk scenario as the technologies exist for insurgents to utilize UAVs in this way, this is much more likely a near futures issue that could still be some years out on the horizon before non-state opposing forces even contemplate or attempt such attacks.

Since no terrorism component is readily foreseen but rather force-on-force engagements are being focused upon, this is not viewed as a domestic law enforcement and homeland security concern. Rather, it is an Army and allied services expeditionary concern bridging the tactical into the operational level of impact.

Strategic.

While the drone swarms of normal and micro-sized UAVs projected in this threat scenario may still be a few decades out and possibly even beyond the capacity of terrorist and insurgent forces to field on their own without state sponsorship, now is the time to attempt to get ahead of such developments and help shape the future combat environment. At a minimum, we may presently be in an inter-war period, as experienced between World Wars I and II, when the various elements leading to a Revolution in Military Affairs took place with the evolution of the tank and supporting arms that resulted in the mass armor and mechanized formations that fought in World War II.

In this instance, similar disparate elements, involving robotics, expert systems, artificial and collective (cloud-like based) intelligence, network communications, and 3rd and 4th dimension (3D and 4D) replicators may be ushering in another revolution in land warfare involving both ground and aerial based unmanned vehicles and systems.

UAVs have increasingly been in the news as the cost of these systems continues to drop and their popularity increases. A few million of these systems are now said to exist globally, with the numbers rapidly increasing. Recent stories highlighting UAVs have caused quite a bit of sensationalism and have been focused on increasing concerns over their terrorism potentials. Interestingly, serious concerns over terrorist use of UAVs, and later insurgent use of UAVs (that includes terrorism as an insurgent tactic), have existed for roughly over a decade, but such concerns had not been widely disseminated by the media until recently.

The U.S. Army and the rest of the U.S. governmental defense community have a vested interest in better understanding this area of threat concerns and potentials. While terrorist and insurgent use (and projected use) of UAVs is important for its homeland defense and defense support of civil authorities (DSCA) implications, it is also—and quite possibly more importantly—likely to have great influence on the conduct of future forms of conventional warfighting. The reason for this contention is because, ultimately, UAVs represent artifacts belonging to the ongoing informational and robotics revolutions that have been taking place for decades. The significance of advances in information systems and robotics and what this will mean to future warfighting have not been lost on the Strategic Studies Institute of the U.S. Army War College or on other U.S. defense policy institutions.

With these thoughts in mind, this monograph will provide context related to a short UAV historical overview and their present use by the U.S. military, a section on terrorist and insurgent use (and attempted use) of UAVs, UAV baselines and trending analysis, potentials based on projected UAV threat scenarios, what this may mean in terms of U.S. military implications, and finally suggested forms of policy response at the tactical, operational, and strategic levels.

The first military use of UAVs dates back to World War I when early tests were carried out—one in 1917 by the United Kingdom (UK) involved a radio-controlled Sopwith Camel biplane loaded with dynamite. In World War II, about 15,000 UAVs were built in one Southern California plant alone for anti-aircraft targeting purposes. U.S. military interest and use of UAVs waxed and waned during the Cold War. UAV use then drastically increased due to Section 220 of the National Defense Authorization Act, FY 2001 (from 2000) which mandated the fielding of unmanned air and ground vehicles, combined with the September 11, 2001, attacks on the World Trade Towers and the Pentagon, resulting in the use of armed drones in increasing numbers in the global war against al-Qaida.

Terrorist and insurgent use (and attempted use) of UAVs spans the 1994 Aum Shinrikyo cult's attempt to use weaponized drones through the 2015 Islamic State (IS) use of these craft for reconnaissance and propaganda video purposes. Such groups are still very much in an experimental phase of using these craft and possess relatively few of them, and—when they do have them in their inventories—they tend to be inferior commercial models (as opposed to military grade UAVs). Still, their use by terrorist and insurgent groups is increasing, as are the capabilities of the systems being deployed. During that time span, al-Qaida, the Revolutionary Armed Forces of Colombia—People's Army, Fatah, Hizbollah, Lashkar-e-Taiba, and Hamas have all been involved in actual or attempted UAV use. The purpose of this use has included reconnaissance and surveillance, messaging, IED delivery, weapons of mass destruction delivery, and as a weapons platform. Other UAV capabilities that exist—yet have not been tied to terrorist or insurgent use so far—are smuggling, limited electronic intelligence capability, logistical resupply, and surrender of opposing force personnel. Recent technology trends that may influence future nonstate threat potentials are smart glasses and virtual reality goggles, apps and modular payloads, expert systems and artificial intelligence, and three-dimensional (3D) printing.

Transitioning from present baselines of terrorist and insurgent use of UAVs, along with technology trends influencing their potential uses, three red teaming threat scenarios have been created for early warning purposes: 1) Single UAV—human controlled with drone-up shooting (like a walk-up shooting), IED crowd targets, and aircraft takedown variants, 2) Groups of UAVs—human controlled or semi-autonomous with squad-sized virtual martyr units and semi-autonomous drone squadron variants; and 3) Swarms of UAVs—considered as autonomous to highlight the projected evolution of this weaponry use with drone swarm and micro-drone swarm variants.

These three threat scenarios result in three corresponding levels of impact found at the tactical, operational, and strategic levels of military significance. For the U.S. Army, the tactical implications of such UAV use will fall within force protection, counterterrorism, and defense support of civil authorities' missions. It will focus on UAV detection, countermeasures, and tactical response and is an immediate concern. The operational level of impact is insurgency environment focused and most likely a near futures issue. It pertains to the use of groups of human controlled and semi-autonomous UAVs and represents an expeditionary concern bridging the tactical into the operational level of impact. This means that experimentation and red teaming is warranted related to threat forces' use of UAVs in insurgency type environments. The strategic level of concern, on the other hand, may still be a few decades out, and possibly even beyond the capacity of terrorist and insurgent forces to field on their own without state sponsorship. Still, its autonomous and semi-sentient drone swarm potentials are viewed as having an immense impact on the future conduct of war. Considerations need to be made concerning arms control regimes related to such autonomous, intelligent, and lethal robotic systems as well as their integration with human soldiers into future force structures, if that Army unit composition is elected to be followed—which presently appears to be the national trajectory.

TERRORIST AND INSURGENT UNMANNED AERIAL VEHICLES: USE, POTENTIALS, AND MILITARY IMPLICATIONS

Unmanned aerial vehicles (UAVs), sometimes called simply "drones," have increasingly been in the news as the cost of these systems continues to drop and their popularity increases.[1] At present, a few million of these systems are said to exist globally, with the numbers rapidly increasing.[2] Recent stories highlighting UAVs have caused quite a bit of sensationalism and have been focused on increasing concerns over their terrorism potentials. Such stories have included:

- In London, a December 2014 governmental confirmation of an earlier incident of a UAV almost hitting an airliner at Heathrow Airport followed by ongoing illegal drone flights over city landmarks and sports stadiums in 2015.[3] These incidents have increasingly sensitized the public and officials to drone terrorism threat potentials—even including those to United Kingdom (UK) nuclear plants.[4]
- A small quadcopter that penetrated Secret Service security and crash-landed on the White House lawn on January 27, 2015. It turns out the UAV was being flown for recreational purposes by a U.S. Government employee around midnight who said that he lost control of the device (known as a flyaway).[5]
- Mystery UAVs flying over Paris in late-February and early-March 2015 then created a mini-hysteria in a city already on edge from the earlier mid-January Charlie Hebdo, supermarket, and printing firm gun battles involving radical Islamist terrorists proclaiming allegiance to

al-Qaida and the Islamic State (IS).[6] While the UAV intrusions were shown to have no terrorism links, the psychic damage had already been done—much of the French public now realize that terrorists could use drones for future attacks.

Interestingly, serious concerns over terrorist use of UAVs, and later insurgent use of UAVs (that includes terrorism as an insurgent tactic), have existed for roughly over a decade but such concerns had not been widely disseminated by the media until recently.[7] The U.S. Army and the rest of the U.S. governmental defense community have a vested interest in better understanding this area of threat concerns and potentials—not because they have now been sensationalized but because of the underlying early warning patterns that they have been generating for some time.

While terrorist and insurgent use (and projected use) of UAVs is important for the Army's homeland defense and defense support of civil authorities (DSCA) implications, it is also—and quite possibly more importantly—likely to have great influence on the conduct of future forms of conventional warfighting. The reason this contention is being made is because ultimately UAVs represent artifacts belonging to the ongoing informational and robotics revolutions that have been taking place for decades.[8] Such artifacts, when utilized for conflict and war, are, of course, not only being employed by violent nonstate actors but also by sovereign states. States, indeed, have almost totally monopolized this combat capability until quite recently. In fact, the United States and its allies have been without peer in their utilization of UAVs since September 11, 2001 (9/11), to target and engage a number of the terrorist and insurgent groups identified in this monograph and even, at times, using UAVs against targets associated with the sovereign states which sometimes harbor these terrorists or insurgents.

The significance of advances in information systems and robotics and what this will mean to future warfighting have not been lost on the Strategic Studies Institute of the U.S. Army War College or on other U.S. defense policy institutions.[9] In December 2015, Dr. Steven Metz wrote a short yet prescient essay about the coming Landpower robot revolution and provided five questions related to its first phase based on early innovation and experimentation:

1. What is the appropriate mix of humans and robots?
2. How autonomous should the robots be?
3. What type of people will be needed for robot heavy Landpower formations?
4. What effect will robot centric Landpower have on American national security policy?
5. What to do about enemy robots?[10]

This monograph addresses some of Metz's questions in the areas of projected enemy (terrorist and insurgent) UAV (robot) capabilities and their level of autonomy. Also, some mention of envisioned threat forces of mixed humans and robots will be highlighted. While this discussion will not specifically provide guidance related to future American robot Landpower, it may help to provide some analytical preconditions for such an effort. With these thoughts in mind, this monograph will provide context related to a short UAV overview and their present use by the U.S. military, a section on terrorist and insurgent use (and attempted use) of UAVs, UAV baselines and trending analysis, potentials based on projected UAV threat scenarios, what this may mean in terms of U.S. military implications, and finally suggested forms of policy response at the tactical, operational, and strategic levels.

UAV OVERVIEW
PRESENT U.S. MILITARY USE

The first military use of UAVs dates back to World War I when early tests were carried out—one test in 1917 by the UK involved taking a radio-controlled Sopwith Camel biplane loaded with dynamite to see if it could be made to ram into one of the German zeppelins, the craft that were then bombing British cities. The demonstration had to be scrapped due to radio command and control failures which almost resulted in a group of gathered generals on the ground being killed by what had then become a rogue UAV diving towards them.[11] Testing continued on and off by the United States, the UK, and others over the ensuing years with drone use for anti-aircraft targeting practice becoming common in the 1930s.

In World War II, about 15,000 UAVs were built in one Southern California plant alone for such purposes.[12] Attempts at creating unmanned B-17 and B-24 bombers, which were conceptually based on the earlier World War I Kettering Bug concept—to dive into highly defended German military-industrial targets—were also attempted. These drones were beset with numerous issues—including the fact that they required human operators to get them airborne and arm their explosive charges prior to bailing out from the planes—and achieved very limited results. The Germans utilized drones in a different way, with 8,000 of their infamous V-1 flying bombs being launched later in the war against Britain in an indiscriminate terror campaign.[13]

U.S. military interest and use of UAVs waxed and waned during the Cold War. Reconnaissance drones were used in Southeast Asia, based on an initial 1962 contract, with over 3,000 missions of the Fire Fly crafts flown. Then, between 1979-87, the failed Army Aquila project—which sought to create 780 drones that could relay operational level battlefield intelligence—resulted in only a few prototypes being produced at a cost of over $1 billion. One of the limited drone successes during this era can be attributed to Israel. In its 1982 Bekaa Valley attack on Syrian air defenses (situated in Lebanon), an initial wave of UAVs triggered the system, which proceeded to fire its missiles at the decoy drones. While the Syrians were in the process of reloading their own missiles, a second wave of Israeli jets came in and fired their radar homing missiles, wiping out the Soviet derived air defense system.[14]

Until 9/11, U.S. military drone use existed at a low yet somewhat steady level, with some of the older Fire Fly (renamed Lightning Bug) units still in existence along with the newer RQ-2 Pioneer system fielded in 1986 and considered vital for battlefield reconnaissance (and later targeting) missions by the various services. Limited chaff and propaganda (leaflet) dropping missions also took place with some of these UAVs. Further, the Central Intelligence Agency (CIA) quietly became involved in influencing UAV fielding and use from the 1980s onward, with the emergence of their large Predator surveillance drone (and its dedicated satellite links) in 1994, which was deployed to the Balkans in the mid- and late-1990s.[15]

Section 220 of the *National Defense Authorization Act, FY 2001* (from 2000) established the following initial goal—then changed the equation:

(a) Goal.—It shall be a goal of the Armed Forces to achieve the fielding of unmanned, remotely controlled technology such that—

(1) by 2010, one-third of aircraft in the operational deep strike force aircraft fleet are unmanned; and,

(2) by 2015, one-third of the operational ground combat vehicles are unmanned.[16]

When combined with the 9/11 attacks on the World Trade Towers and the Pentagon, these events not only resulted in a firebreak promoting future mandated UAV use by military forces but saw those systems, along with UAVs belonging to the CIA, used in increasing numbers in the global war against al-Qaida. Prior to those dual events, only small numbers of UAVs (estimated at less than 50) were being utilized by the intelligence community and armed services. About a decade later, a 2012 Congressional Research Services (CRS) report entitled *U.S. Unmanned Aerial Systems* (UAS) identified 7,494 Department of Defense (DoD) UAS platforms in the inventory.[17] This number did not include CIA dedicated units, which conservatively are estimated at 30 but could be somewhat higher.[18] In the CRS report, UAVs now equal about 70 percent of the manned U.S. aircraft inventory (which stands at 10,767) and are collectively engaged in all of the following DoD capabilities and missions:

- Anti-Submarine Warfare
- Anti-Surface Warfare
- Battle Management Command and Control
- Electronic Warfare

- Explosive Ordnance Disposal
- Force Protection
- Intelligence, Surveillance, and Reconnaissance
- Maritime Domain Awareness
- Mine Warfare (Naval)
- Organic Mine Countermeasures (Naval)
- Precision Strike
- Reconnaissance, Surveillance, and Target Acquisition.[19]

As can be seen, the U.S. military is now heavily invested in UAVs for numerous warfighting capabilities with over 70 years of learning experience associated with their fielding and use. Sovereign state militaries—both allies and potential belligerents—increasingly are now deploying these systems in an attempt to catch up to the superior U.S. capabilities in this technological area. In the following section on terrorist and insurgent use of UAVs, however, it can be seen that their fielding of these devices is much more of a haphazard and limited affair. These groups are still very much in an experimental phase of using these craft and possess relatively few of them, and—when they do have them in their inventories—they tend to be inferior commercial models (as opposed to military grade UAVs). Still, their drone use is now increasing as are the capabilities of the systems being deployed.

Terrorist and Insurgent Use of UAVs.

The use, and attempted use, of UAVs by terrorists and insurgents can at least be dated back to the pre-June 1994 attempts by the Japanese apocalyptic cult Aum Shinrikyo to conduct dry runs to release the nerve agent sarin by means of remote controlled helicopters with aerial spray systems.[20] The attempts failed as the mini-helicopters crashed during testing, with that terrorist group going on to utilize different dispersal methods when they launched their sarin attacks on a Matasumoto courthouse and later on the Tokyo subway system.[21] The latter attack resulted in about a dozen people killed and 5,500 injured by this nerve agent.

The next incident related to UAV threatened use was that of a pre-July 2001 improvised explosive device (IED) attack upon G8 Summit leaders (Canada, France, Germany, Italy, Japan, Russia, the UK, and the United States) in Genoa, Italy, by al-Qaida. This plot may have only entered the "what if" stage of conceptualization with Osama bin Laden musing about its potentials as discrepancies exist as to whether the plot was actually ever put into place.[22] Two more al-Qaida based plots followed: one pre-February 2002 originating out of Pakistan, and the other in June 2002 from an unspecified location. The first plot tied to Mozzam Begg sought to launch a drone filled with anthrax against the English House of Commons. He was sent to Guantanamo Prison for his involvement but was later released from custody in January 2005 because the original charges became questionable.[23] The second plot revolved around IED-carrying remote controlled planes being utilized against passenger airlines—though the plot was never said to get beyond the concept stage.[24]

Then, in August 2002, a Colombian Army unit seized nine remote controlled planes from a camp deep in the jungle belonging to the Fuerzas Armadas Revolucionarias de Colombia (FARC) guerrilla group. The intended use of these planes is unknown, but some speculation existed that they may have been intended to carry IEDs.[25] This was followed by a Fatah Palestinian plot in December 2002 to conduct IED model airplane attacks on Jewish sections of Jerusalem. The plot, which was to involve hundreds of such model airplanes, never got beyond the flight test stage.[26]

Three linked incidents in which UAVs were to be utilized for attack, reconnaissance, and protest purposes subsequently took place. The initial one from August through December 2003 was a joint effort by a Hizbollah cell that was supporting the Al Aqsa Martyrs brigade, an arm of Fatah. The intent was to launch an IED UAV attack on Jewish settlers in Gaza, however, the plot was interdicted by Israeli security forces.[27] Then, on November 7, 2004, a Hizbollah drone was launched from southern Lebanon and engaged in a 20-minute reconnaissance over Nahariya in northern Israel. The Mirsad-1 drone, provided by Iran, was of military grade quality with conflicting reports of its either crashing in the sea off the Lebanese coast or returning back to its Hizbollah base after its reconnaissance flight.[28] The final incident took place on April 11, 2005, and involved another Mirsad-1 drone flown by Hizbollah from southern Lebanon. In this incident, the drone overflew the northern Israeli city of Acre as a protest of Israeli airspace violations of Lebanon, according to Hizbollah. The drone was able to complete its mission successfully and return back to its Hizbollah base.[29]

Two Pakistani terrorist group linked incidents then took place on September 13-14, 2005. In the first, the Pakistani Army raided an al-Qaida hideout in North Waziristan. In the raid, they seized a Chinese made remote control model airplane which was said to be used for the reconnaissance of Pakistani security forces prior to attacking them. IED weaponization potentials of this model aircraft were also mentioned.[30] The next day, Ala Asad Chandia (Abu Qatada) was arrested in Fairfax County, VA, for obtaining an MP 1OOOSYS electronic automatic pilot system for model aircraft. This Lashkar-e-Taiba trained individual was federally indicted and subsequently convicted for attempting to send this technology to that terrorist group for its drone use in Pakistan.[31]

In what may or may not be considered a major escalation of terrorist and insurgent drone capabilities, on August 13, 2006—during the Second Lebanon War—three Ababil (military grade) Iranian drones supplied to Hizbollah were launched against Israel from southern Lebanon. Each drone was said to be carrying a 40-50 kilogram explosive warhead and was intended for use against a "strategic target," according to Hizbollah. The threat was taken seriously enough that F-16 Israeli fighters shot down these UAVs near Tyre in Lebanon and near Haifa and Western Galilee in Israel. Upon inspecting the wreckage of some of these craft, Israel claimed that they were not carrying warheads.[32]

Between 2006 and May 2012, two al-Qaida incidents and one Taliban UAV incident occurred. In Columbus, OH, during the 2006-07 period (exact dates unspecified), al-Qaida trained Christopher Paul was conducting drone research, utilizing a 5-foot-long model helicopter, for terrorism purposes. He was arrested by the Federal Bureau of Investigation (FBI) in August 2007 and was subsequently convicted in June 2008 (he took a plea deal that resulted in a shorter sentence).[33] In the next incident, which took place on September 28, 2008, Rezwan Ferdaus—an al-Qaida affinity adherent—was arrested by the FBI in Ashland, MA. He was caught in a terrorist sting operation related to his plot to drive F-86 Sabre and F-14 Phantom scale models (with Global Positioning System [GPS] capability) loaded with C-4 explosives into the Pentagon and Capitol buildings. He was convicted for this plot in 2012 and also took a plea deal, like Paul, for shorter sentencing purposes.[34] Finally, on May 19, 2012, an allied raid on a Taliban base in Helmand Province, Afghanistan, turned up a small drone—possibly a North Atlantic Treaty Organization (NATO) Desert Hawk—along with some IEDs and small arms. The intended Taliban use of the drone was unknown, quite possibly for reconnaissance purposes, though it appeared slightly damaged in a photo of the arms cache and no control unit was found along with it.[35]

Four more escalatory Hizbollah and Hamas UAV incidents took place between the latter part of 2012 and mid-2014. On October 6, 2012, Hizbollah sent an Iranian Ayoub drone over Dimona, Israel—a restricted area which contains that nation's nuclear weapons facilities—for reconnaissance purposes. The timing of the drone incident coincided with Israeli military exercise preparations. Given the sensitivity of this area, the drone was shot down by an Israeli F-16, although not until after it had been aloft for some hours.[36] Another Hizbollah drone, the type not specified, was then shot down by an F-16 10 kilometers out to sea west of Haifa on April 22, 2013. What mission this UAV was engaging in is unknown.[37] A Hamas plot at a local university to send a UAV carrying explosives into Israel in October 2013 was then interdicted by the Palestinian Authority in Hebron.[38] In the last of these incidents, on July 14, 2014, a homemade Hamas drone was shot down over Ashdod, Israel, by a patriot missile. This 5-foot-long drone was outfitted with small air-to-ground rockets (per unconfirmed Hamas video images) and was on its way to engage an undisclosed Israeli target.[39]

The Islamic State (IS) joined the nonstate threat drone proliferation club with three successful UAV operations in August and September 2014. The first incident was on August 23, 2014, near Raqqa province in northern Syria. It involved the use of a commercial system—a DJI Phantom FC40 quadcopter—to recon Syrian Army Base 93 prior to an IS ground assault on the base. The quadcopter video imagery was subsequently used in IS propaganda videos.[40] On August 30, 2014, an unspecified IS drone was used over Falluja, Iraq, to provide imagery of attacks on the city for online propaganda purposes.[41] In the last IS operation on September 12, 2014, in Kobani, northern Syria, another unspecified drone was used to capture video imagery of suicide bomber and ground attacks on that city for propaganda purposes.[42]

Hizbollah then engaged in a successful drone strike operation against the al-Nusra Front—an al-Qaida linked group—near Arsal in northeastern Lebanon on September 21, 2014. Twenty-three al-Nusra terrorists were said to be killed in this attack, which was followed up by a group assault—an incident that has now ushered in terrorist-on-terrorist group based drone warfare.[43] Whether the drone utilized in the attack carried an explosively tipped warhead or carried air-to-ground rockets (or missiles) is unknown.[44] A final incident involving terrorist and insurgent use of UAVs occurred on around March 16, 2015, near the city of Fallujah, Iraq. In that incident, an IS militant flew a small model aircraft for about 20 minutes. After the drone landed, the IS operative placed the drone in the trunk of a car and proceeded to drive off, at which point U.S. coalition military forces launched an airstrike destroying the insurgent, the drone, and the vehicle.[45] This incident and all of the earlier ones summarized in this section can be viewed in Table 1.[46]

Date	Location	Perpetrator	UAV Type	Use	Outcome
Pre-June 1994	Japan	Aum Shinrikyo	Remote Controlled Helicopters	Spray Chemical Agent (Sarin)	Crashed During Testing
Pre-July 2001	Genoa, Italy	Osama bin Laden, al-Qaida Leader	Remote Controlled Airplanes	IED Attack on G8 Summit Leaders	Considered Only; Not Attempted (Alleged)
Pre-February 2002	Pakistan	Moazzam Begg, al-Qaida Operative	Drone	Launch Drone from Suffolk with Anthrax Against House of Commons	Alleged Plot; Sent to Guantanamo Prison; Released in January 2005
June 2002	Not Specified	al-Qaida	Remote Controlled Airplanes	IED Attack on Passenger Airliners	Considered Only; Not Attempted (Alleged)
August 2002	Colombia	FARC	9 Remote Controlled Airplanes	Unknown; Possibly Weaponized (IED)	Recovered by Colombian Army Unit from Remote Camp
December 2002	Jerusalem, Israel	Fatah	Hundreds of Model Airplanes	IED Attacks on Jewish Sections of Jerusalem	Conducted Flight Tests Only
August and December 2003	Gaza, Palestine	Hizbollah Cell (Linked to Al Aqsa Martyrs Brigades; Fatah)	UAV	IED Attack on Jewish Settlers in Gaza	Interdicted by Israeli Security Forces

Major Chronological Sources: Michael Gips, "A Remote Threat," *Security Management Online*, October 2002; Eugene Miasnikov, *Threat of Terrorism Using Unmanned Aerial Vehicles: Technical Aspects*, Moscow, Russia: Center for Arms Control, Energy and Environmental Studies at Moscow Institute of Physics and Technology, June 2004, translated into English March 2005, available from *www.armscontrol.ru/uav/report.htm*; Jay Mandelbaum and James Ralston *et al.*, *Terrorist Use of Improvised or Commercially Available Precision-Guided UAVs at Stand-Off Ranges: An Approach for Formulating Mitigation Considerations*, ADA460419, Alexandria, VA: Institute for Defense Analysis, October 2005, available from *oai.dtic.mil/oai/oai?verb=getRecord&metadataPrefix=html&identifier=ADA460419*; Milton Hoenig, "Hezbollah and the Use of Drones as a Weapon of Terrorism," *Public Interest Report*, Vol. 67, No. 2, Spring 2014, available from *www.fas.org/pir-pubs/hezbollah-use-drones-weapon-terrorism/*.

Table 1. Terrorist and Insurgent Use/Attempted Use of UAVs.

7 November 2004	Nahariya, Northern Israel	Hizbollah	Iranian Mirsad-1 Drone	20-Minute Reconnaissance Mission	Either a) Crashed in the Sea Near Lebanese Shore or b) Returned Back to Hizbollah Base
11 April 2005	Acre, Northern Israel	Hizbollah	Iranian Mirsad-1 Drone	Overflight of Israeli Communities (Stated as a Protest of Lebanese Airspace Violations)	Successful Operation; Returned Back to Hizbollah Base
13 September 2005	North Waziristan, Pakistan	al-Qaida	Chinese Made Remote Control Model Airplane	To Recon Pakistani Security Forces Prior to Attack; Also Weaponized (IED)	Seized in Major Raid of al-Qaida Hide Out by the Pakistani Army
14 September 2005	Fair Fax County, Virginia	Ala Asad Chandia (Abu Qatada); Lashkar-e- Taiba Trained	Obtained MP 1000SYS—Electronic Automatic Pilot System for Model Aircraft in April 2002	For Lashkar-e-Taiba Terrorist Group Drone Use in Pakistan	Indicted and Subsequently Convicted
13 August 2006	Near Tyre, Lebanon; Near Haifa, Israel; Western Galilee, Israel	Hizbollah	3 Ababil Drones, Each With 40-50 Kilogram Warhead	Against "Strategic Targets"	All 3 Shot Down by Israeli F-16s
2006-2007	Colombus, Ohio	Christopher Paul (al-Qaida trained)	5-Foot-Long Model Helicopter	Conducted Drone Research for Terrorism Purposes	Arrested by FBI and Convicted (Plea in 2008)
28 September 2011	Ashland, Massachusetts	Rezwan Ferdaus (al-Qaida Affinity)	Scale Models of F-86 Sabre and F-14 Phantom Jets (GPS capability)	IED (C-4 Explosive) Attack on Pentagon and Capitol Buildings	Arrested By FBI and Convicted (Plea in 2012); Sting Operation
19 May 2012	Helmand Province, Afghanistan	Taliban	Might be Recovered NATO UAS (Desert Hawk Drone)	Unknown Use - Possible Recon. Found with IED Materials and Small Arms	Captured in Raid
6 October 2012	Dimona, Israel	Hizbollah	Iranian Ayoub Drone	Recon of Israeli Nuclear Weapons Complex & Military Exercise Preparation	Shot Down by Israeli F-16 Jet

Table 1. Terrorist and Insurgent Use/Attempted Use of UAVs. (cont.)

22 April 2013	Over the Sea 10 Kilometers West of Haifa, Israel	Hizbollah	Unmanned Drone; Type Not Specified	Mission Unknown	Shot Down by Israeli F-16 Jet at an Altitude of 6,000 Feet
October 2013	West Bank, Palestine	Hamas	Plot Centered at Hebron University to Place Explosives on UAV	Fly into Israel to Engage Unknown Target(s)	Palestinian Authority Arrested Plotters Prior to Launch
14 July 2014	Ashdod, Israel	Hamas	5-Foot-Long Homemade Drone Aircraft with Small Rockets (Unconfirmed Hamas Video Image)	Fly into Israel to Engage Unknown Target(s)	Shot Down by Israeli Patriot Missile
23 August 2014	Near Raqqa Province, Northern Syria	Islamic State (IS)	DJI Phantom FC40 Quadcopter	Recon of Syrian Army Military Base 93 Prior to Ground Assault; Imagery Provided via IS Propaganda Video on YouTube	Successful Operation
30 August 2014	Falluja, Iraq	Islamic State (IS)	Unspecified Drone	Propaganda Purposes; Video of Attacks in the City	Successful Operation
12 September 2014	Kobani, Northern Syria	Islamic State (IS)	Unspecified Drone	Propaganda Purposes; Video Footage of Suicide and Ground Attacks	Successful Operation
21 September 2014	Near Arsal, Northeastern Lebanon	Hizbollah	Armed Drones	Killed 23 al-Nusra Front (al-Qaida Linked) Fighters at Base; Followed by Ground Assault	Successful Operation
Appx. 16 March 2015	Near Fallujah, Iraq	Islamic State (IS)	Unspecified Drone	Unknown; Possible Reconnaissance or Propaganda Purposes	Operator and Drone Destroyed in Car by U.S. Coalition Air Strike

Table 1. Terrorist and Insurgent Use/Attempted Use of UAVs. (cont.)

BASELINES AND TRENDING ANALYSIS

Derived from the overview of the historical and contemporary use (and attempted use in plots) of UAVs by terrorists and insurgents in the preceding section, the following capabilities gained by using this technology have been identified.

Reconnaissance and Surveillance.

Initially, the reconnaissance of fixed facilities and military units was gained by UAV deployment, as was evident with Hizbollah drone use in November 2004 and October 2012 against Israel. Also, an al-Qaida drone was seized in September 2005 in Pakistan which would be used prior to launching an attack (possible use) as was a Taliban drone seized in May 2012 in Afghanistan (possible use). An IS drone was also used in a reconnaissance role as recently as August 2014 against a Syrian army base in northern Syria prior to a ground assault upon it. The use of drones for real time surveillance appears to be a far less common UAV occurrence for violent nonstate actors, although real time drone imagery of IS attacks on Falluja, Iraq, in August 2014 and Kobani, Syria, in September 2014 used for propaganda purposes could conceivably also be used for command, control, and coordination purposes—but, in those examples, this was probably not exploited.[47]

Messaging.

Various forms of messaging (communicating information to others) exist related to UAV use. At the most basic level, this can take the form of making a **protest**. It was thought that the April 2005 Hizbollah use of a drone to overfly Israeli communities was primarily meant to signal that an earlier Israeli violation of Hizbollah airspace in southern Lebanon would not be tolerated.[48] Another form of messaging is that of **propaganda**. Propaganda can be directed at both internal and external audiences in terms of drone use capability and actual use. The use of such propaganda has been readily capitalized on by al-Qaida, Hizbollah, Hamas, and IS in their online postings and videos. A third form of messaging is to give a **warning**. This is very much akin to the old "shot across the bow," which means that deadly force will likely be used next unless the targeted audience accedes to one's requested demands. Providing a warning to the other side in such a manner may also serve as a form of future deterrence against unwanted actions. The October 2012 Hizbollah drone reconnaissance of the Dimona, Israel, nuclear weapons complex as well as the propaganda component of the July 2014 Hamas drone incident (both physical and online) in Ashdod, Israel, that "we now, too, have armed drones" were meant to threaten Israel for deterrence and behavioral shaping purposes.

IED Delivery.

Since pre-July 2001, al-Qaida leaders have been musing about using drones equipped with IEDs for terrorist attack purposes against Western leaders. Al-Qaida plots via affinity nodes have also included targeting passenger airliners (June 2002), general testing (2006-07), and use against key governmental buildings in Washington, DC (September 2011). A Fatah plot (December 2002) and one intertwined with Hizbollah (in August and December 2003), along with an actual Hizbollah incident involving explosive payload drones (August 2006) shot down before hitting their targets, have also been identified. More recently, in September 2014, a Hizbollah drone was used to attack the al-Nusra Front (an al-Qaida linked group) near Arsal in northeastern Lebanon (assumed on the Syrian side of the border). In this successful operation—quite possibly the first of its kind for a nonstate group—some 23 al-Nusra fighters in a base were said to be killed in the attack that was then followed up by a ground assault. It is unknown if the drone was carrying an explosive payload or utilized an air-to-ground weapon to destroy the command facility the al-Nusra personnel occupied.[49]

Weapons of Mass Destruction (WMD) Delivery.

The pre-June 1994 attempt by Aum Shinrikyo to weaponize a UAV to spray the sarin nerve agent, and the alleged pre-February 2002 plot by al-Qaida operatives to release anthrax against the House of Commons in London, UK, suggest that this potential drone capability has long been identified by terrorist groups. The delivery of radiological materials by means of a UAV would represent another component of such WMD capability, though it has not been linked to any known terrorist plots. Still, such drone WMD use potentials are widely recognized by security analysts: "Drones could potentially carry and launch some weapons of mass destruction—biological and chemical weapons and even radioactive 'dirty' bombs."[50]

Weapons Platform.

Placing rockets and missiles on drones (to mimic far more robust U.S. drone capabilities) represents another violent nonstate actor capability that can be derived from UAV employment. This trend appears to be relatively recent and may be evident in the homemade Hamas drone outfitted with small rockets under its wings in the July 2014 Ashdod incident. This capability may already be possessed by Hizbollah via the possible transfer of the armed "Hamaseh" Iranian produced drone. Since Hizbollah has already utilized a number of other Iranian drone types in the past, it takes little imagination to see this new drone ending up in Hizbollah's armory. A photo of this drone was taken in May 2013 and serves as a weapons platform for two air-to-ground attack munitions.[51] Another more basic capability—that of placing a firearm on a UAV and using it to shoot at a target—has already taken place. This can be seen in an online advertisement for smart phone shields in which a pistol attached to a drone fires at various items, including a smart phone utilizing the advertised product.[52] To date, the placing of firearms on UAVs has not been tied to any known terrorist or insurgent plots or incidents, but the potentials are being discussed in online media.[53]

Other UAV Capabilities.

Other UAV capabilities presently existing that have not been tied to terrorist and insurgent use or plots are:

Smuggling: Since at least 2009, numerous examples exist globally of UAVs being used by criminals and organized crime to smuggle goods such as narcotics, cigarettes, and cell phones into fixed installations such as prisons, and even across national boundaries.[54] Reports suggest that one of the major regions in which such smuggling has taken place is along the U.S. and Mexican border. It is being carried out by the Mexican cartels, with well over 100 incidents said to have taken place.[55] This UAV capability would allow for the raising of revenues via narcotics smuggling by terrorists and insurgents.

ELINT Capability (Limited): In 2014, a drone was created to hack mobile devices specifically by means of finding those with open Wi-Fi network connections and tricking them into providing data by mimicking networks they have accessed in the past. This was done by equipping the drone with the Snoopy software capability.[56] This form of ELINT drone can be used for illicit fundraising (eg., bank account access), to engage in identity theft in order to compromise cyber and physical security systems, and for gaining intelligence against individuals for kidnapping purposes.

Logistical Resupply: The first UAV combat resupply took place in December 2011 when a K-Max helicopter adapted by Lockheed brought in supplies to a Marine base in Afghanistan.[57] This program has since continued and now progressed to where a software application (app) is being developed to allow Marines to summon resupply UAVs via tablets and smart handheld devices. The experimental program is funded through 2018 and, if successful, we could see more widespread introduction of this system in the 2020s.[58] Such a basic UAV resupply capability would mean that frontline advancing or besieged insurgent fighters could also potentially benefit from food and ammunition supplies being flown into them via modified commercial UAV systems.

Surrender of Opposing Force Personnel: During the Gulf War, in late-February 1991, a small group of Iraqi troops surrendered to a U.S. Pioneer RQ-2A drone on Faylaka Island near Kuwait City rather than face another bombardment of 16-inch shells from the U.S.S. *Missouri*. The low flying drone was being used as a spotter via a video link back to the *Missouri* for targeting and battlefield damage assessment of the defending Iraqi forces.[59] This was the first time in history that a group of soldiers have ever been known to surrender to a robot in war. While a terrorist group may or may not find such a UAV capability useful, it may have utility for an insurgent group that is attempting to capture a city.

Recent technology trends that may bolster UAV functionality suggest that the following enabling technologies may also influence future terrorist and insurgent potential uses.

Smart Glasses and Virtual Reality Goggles: UAVs are typically flown using handheld controllers like those used for model airplanes, with the pilot observing the drone from a distance. This form of stand-off piloting is functional when the UAV is being flown in noncomplex terrain such as in open fields, and to a limited extent over urban areas, when pilot line-of-sight is maintained. An immediate drawback to this method of UAV control is pilot perspective—they are viewing the drone from afar—which reduces its tactical maneuverability and handling. To overcome this limitation, hobbyist and commercial UAV pilots have taken to using smart glasses (with see-through lenses that have computer imagery projected on them) and virtual reality goggles and visors that create computer-generated three-dimensional (3D) simulations of what a drone sees in front of it via a video camera system.[60] This merges the perspective of the stand-off pilot with the UAV—in essence, placing him in the drone or allowing him to become the drone (as a "virtual martyr")[61] for flight control purposes. The end result is that tactical mobility greatly increases as can been seen in a clip of drone races through a forest in Argonay, France, which was posted on *YouTube* in September 2014 and has been widely viewed.[62]

Apps and Modular Payloads: The trend towards open architecture systems—"Plug-N-Play"—will mean that UAVs can be quickly configured for different uses via apps and payload modularity (hardware). One off-the-shelf method in this regard is to attach a smart phone (eg., iPhone or Galaxy) to a UAV in order to gain new forms of functionality for various uses.[63] Novel smart phone apps of interest include GPS fencing which would designate a limited geographic area that a drone could only fly within for patrol or seeking purposes[64] and an infrared video attachment that would allow a drone to be flown under the cover of darkness and also pick up target body heat signatures.[65] Payload modularity means that a drone could transition from cargo hauling through reconnaissance through serving as a weapons platform or as an actual aerial IED as required.

Expert Systems and Artificial Intelligence: One of the limitations of UAV use is that the systems have to be constantly monitored and controlled by human beings. It is expected that expert systems ("if-then statement" decision point) and artificial intelligence (scenario maximizing) drone controllers will also be employed. The mission value is in some ways equivalent to utilizing a wire-guided missile that the operator needs to keep on target as opposed to a fire-and-forget weapon that can be launched and then does not require human interaction to guide it to the target. The ability to send out an autonomous UAV to complete a simple mission already exists—such as for a computer program to fly a drone and take some form of basic action.[66] The use of such semi-independent and independent systems would also get around limitations in controller signal range—extending drone flying distances—and would likely allow for far quicker reaction cycles to changing operating conditions (e.g., machine decisionmaking and flight corrections are faster than that of human beings). Additionally, machine based groups and swarms of drones can operate together in coordinated (intelligent) networks, which is beyond the capability of groups of human controlled UAVs.[67]

3D Printing: Entire UAVs, except for certain motor and command and control parts, can now be created by 3D printers. The first printed drone parts—in this instance, for a model aircraft—were created in 2011 by Southhampton University and took a week to print.[68] By 2014, drone components could be printed in less than 24 hours, as was done by a Sheffield University team.[69] Later that year, a military grade fully autonomous drone (with an Android phone brain), the Razor 3, was printed for the MITRE Corporation, a DoD contractor, with off-the-shelf parts for $2,500 in just over a day by the University of Virginia.[70] In tandem with these developments is the 3D printing of a firearm, which took place in 2013 based on a primitive design (the plastic Liberator) while, in November 2013, a metal M1911 pistol was printed using an industrial 3D printer.[71] A projected capability to regularly print higher strength metal components, in addition to plastic and composite components, is expected once the technology to do so becomes economically feasible.

This will mean that the 3D printed core components of a drone outfitted with various forms of weaponry will at some point become a reality. An urban street culture video linking 3D printed guns and drones, gangsters, narcotics, and violence together should also be noted. It can be found in the video, *Double Bubble Trouble*, released by rap singer Mathangi "Maya" Arulpragasam, known as M.I.A., in her 2013 album, Matangi.[72] The concern, of course, is that these technology linkages are being spread to demographics that may include disenfranchised Western youth susceptible to ongoing terrorist radicalization initiatives.

POTENTIALS—THREAT SCENARIOS

Transitioning from present baselines of terrorist and insurgent use of UAVs, along with technology trends influencing their potential uses, three red teaming threat scenarios have been created for early warning purposes: 1) Single UAV—human controlled, 2) Groups of UAVs—human controlled or semi-autonomous, and 3) Swarms of UAVs—autonomous to highlight the projected evolution of this weaponry use (See Table 2.). Each threat scenario will be discussed, along with the expected time frame in which it may take place and its probable significance to U.S. national security and military operations.

Threat Scenario	Time Period	Description	Significance
1: Single UAV—Human Controlled	Present Day	Tactical action utilized to create a terrorism incident. Scenario variants: Drone-up Shooting, IED Crowd Targeting, and Aircraft Takedown	Tactical (+Terrorism Disruptive Potentials)
2: Group of UAVs—Human Controlled or Semi-autonomous	Present Day Near Futures (Some Years)	Force-on-force engagement in insurgency environment. Scenario variants: Squad-sized Virtual Martyrs Unit and Semi-autonomous Drone Squadron	Operational
3: Swarm of UAVs—Autonomous	Futures (A Few Decades)	Robotic targeting of human personnel, materiel, vehicles, aircraft, and vessels in conflict and war. Scenario variants: Swarms and Micro-Swarms	Strategic

Table 2. Terrorist and Insurgent UAV Use Threat Scenarios.

Threat Scenario 1: Single UAV—Human controlled.

In this threat scenario, a single human controlled UAV is utilized in a tactical action for terrorism purposes. Precedent exists for such an incident derived from past terrorist plots and activities tied to al-Qaida and its wider web of affinity-linked individuals. Such a scenario is presently achievable with current off-the-shelf technologies. Three scenario variants will be provided to showcase the diversity of terrorist attacks that can be carried out via a single human operated UAV.

Drone-up Shooting.

In this scenario variant, a low and slow flying drone is utilized for assassination purposes in order to kill a political leader, general officer, or other very important person. Since a firearm was successfully placed and fired from a small UAV in 2013; the only difference this scenario would require is the engagement of a human rather than an inanimate object. The requirements to conduct this attack would be to determine the route and time frame of the targeted individual in order to send the drone out to the appropriate outside ambush location. A video link can be maintained between the operator and drone, and the simple command of having the trigger of the firearm pulled can be sent digitally. A laser pointer can also be added for accuracy purposes.

IED Crowd Targeting.

The second variant represents an area rather than a point target type of drone attack. As in the previous example, a video link and a simple command—in this instance, that of detonation—can be utilized with a UAV carrying an IED. The intent would be to have the drone fly into a crowd of individuals and detonate among them. This would mimic the effects of a terrorist grenade or IED attack on a grouping of people. An effective use of this form of attack would be to attack crowds in a sports stadium or along a parade route in order to generate panic and create a stampede and/or crowd crush-type situation. Follow-on drones, even if unarmed, could be utilized to create the illusion of a coordinated attack for terror generation purposes.

Aircraft Takedown.

Of the three highlighted variants, this one—targeting a passenger airliner or military jet or transport—could be said to best maximize single human operator

UAV capabilities. Rather than utilizing a quad or other form of slow commercial drone, it would draw upon the speed and kinetic ability of hobbyist scale model jets. The intent of this form of attack is to simulate a "bird strike" on an aircraft engine while an airliner or jet aircraft is taking off and most vulnerable to catastrophic flight failures.[73] Such a UAV strike would be far more serious than a bird strike due to superior kinetic effects. Model jets can achieve speeds well over 100 mile per hour and have more mass than birds[74] and can be augmented with a penetrator rod (composed of a metal or composite material) running the length of a hobbyist jet.[75] As no explosive or form of armament would be required for such an attack—other than a video link for engine targeting purposes—it would be considered a pre-existing off-the-shelf capability.

The significance of this threat scenario (and its variants) ranges from minimal to low in its implications. A drone-up shooting simply represents a variant of a political assassination, while IED crowd targeting mimics a traditional bombing, both of which can be utilized for terrorist (against civilians) and insurgency (against police and military) purposes. The difference, of course, is a standoff between the perpetrator, i.e., the UAV operator, and the UAV being utilized as a weapons platform to carry a firearm or as a delivery system to carry an IED. The flight ability of a UAV, however, may allow it access to venues, such as a sports stadium, where an armed human attacker cannot gain entrance due to security screening protocols.[76] The aircraft takedown variant, however, is different in that a human attacker is precluded from gaining access to an aircraft during takeoff due to both airport and airbase security and the fact the aircraft is in its initial flight stage. This means that utilizing a drone to engage in such an attack represents a new capability that can be directed against civilian airliners for terrorism purposes within the continental United States (CONUS) and against military aircraft for insurgency purposes outside CONUS deployed areas.[77] Ultimately, however, all of these scenario variants leverage the disruptive potentials terrorism has to offer and, while based upon tactical actions, should be considered terrorist attacks first and foremost in their effects.

Threat Scenario 2: Group of UAVs—Human Controlled or Semi-autonomous.

No groups of UAVs have as of yet been utilized together in a conflict setting by violent nonstate actors, such as terrorists or insurgents, so no historical precedent for this threat scenario exists. Rather, it is based on a linear trend projection derived from hobbyist racing for human controlled UAVs and commercial applications derived from various apps being integrated into semi-autonomous drone systems. The first variant, based on the virtual martyrs concept,[78] pertaining to this threat scenario is viewed as a current capability (which may or may not be exploited by Jihadist insurgent groups), while the second variant, highlighting the use of a group of semi-autonomous drones, is expected to become technically feasible in the near future. These scenario variants include the following.

Squad-sized Virtual Martyrs Unit.

This is a squad-sized UAV unit composed of racing drones outfitted with IEDs controlled by virtual reality linked human controllers. The intent of this unit is to attack U.S. and allied soldiers and security personnel by detonating the IED drones once they come into the proximity of their targets. A sub-variant of such drones would be ones outfitted with either shaped charges or explosively formed projectiles (EFP) that would be utilized for the precision targeting of U.S. and allied armored vehicles.[79] Inhibitors to the use of such virtual martyr drones include video feed and controller communication range limitations, signal dead zones, short drone battery life, and IED weight restrictions. This unit can be utilized in a stand-alone mode with only virtually controlled IED carrying drones involved or in a combined arms fashion as ad hoc support to human insurgent fighters and/or integrated with a semi-autonomous drone squadron.

Semi-autonomous Drone Squadron.

Such a squadron represents a small grouping of drones launched together in an assault wave. It would draw upon both the drone-up shooting (firearm carried) and IED crowd targeting (bomb carried) single UAV scenarios, with the addition of autonomous drone capability. The drones can be sent against police and military personnel located at a GPS coordinate. The drones would be provided with GPS fencing instructions to patrol within certain physical boundaries and engage (via weapons firing or IED detonation) humans and/or moving objects that they come across using human form or motion sensors. Human controllers have the option of taking over individual drones for engagement purposes as required. An ambush variant of this scenario variant would be to have these UAVs resting in a "drone nest" (essentially a box buried at ground level) that would open when opposing troops lacking identification friend or foe tags trip sensors during entry into an area.

This scenario and its variants are not about creating terror per se—which may still occur as a secondary outcome—but instead focus on the generation of combat power in force-on-force engagements found in insurgency environments.[80] As a result, the direct implications of this scenario are at the military operational level in which groups of drones serve as a) physical avatars for humans virtually controlling them, or b) machine soldiers controlled by expert systems, both of which directly engage human soldiers in combat. This scenario is thus meant to be reflective of the changing patterns of conflict and war in which the 5th dimension of cyberspace (via physical space-time manipulation)[81] and unmanned systems are beginning to increasingly influence operations.[82] The significance of this scenario would likely be limited in scope initially with increasing importance up to a moderate level of threat represented by the semi-autonomous drone variant when (or if) it is realized.[83]

Threat Scenario 3: Swarm of UAVs—Autonomous.

While this threat scenario may sound like something from a science fiction novel, the technologies that would allow swarms of autonomous drones to operate on the future battlefield are already being developed. One experiment in this regard is being conducted by Hungarian civilian researchers with the COLLMOT Robotic Research Project who have "created 10 drones that self-organize as they move through the air."[84] In one example, these drones overcome an obstacle by maneuvering in line through a choke point. While the researchers strive that this experimentation is for peaceful pursuits, the military benefits of increasingly larger groups of drones that can self-organize in order to fulfill mission requirements is readily apparent. Similar experimentation is ongoing with U.S. Office of Naval Research experiments, which are presently up to 13 autonomous and remote controlled swarming robot boats. As for the desired project counterterrorism end state, "The Navy officials envision swarms of unmanned surface vehicles . . . being used to form a defensive perimeter around larger ships and to surround enemy ships."[85] Based on this projected threat scenario, two variants, utilizing swarms of normal and micro-sized drones, are foreseen.

Drone Swarms.

The full-sized autonomous UAVs incorporated into this scenario variant are meant to destroy major elements of U.S. power such as the military aircraft situated at an air base, the armored vehicles belonging to a brigade combat team, or a major capital ship such as an aircraft carrier. A massive drone swarm—with each semi-intelligent drone networked together and outfitted with an explosive warhead—would be launched against such military targets in this scenario variant. In the case of the targeting of an aircraft carrier, this eventual threat may be considered equivalent in scale to the sinking of captured German battleship *Ostfriesland* in July 1921 by U.S. Army Air Service aircraft under the command of Brigadier General William "Billy" Mitchell. This experiment helped to prove that aircraft carriers would eventually eclipse battleships as the principle capital ship of navy forces and suggests, in the present case of drone swarm potentials, that some sort of stealthy drone carriers may at some point eclipse aircraft carriers as major surface naval combatants.

Micro-drone Swarms.

The micro-drones that form the basis of this threat scenario variant are meant to be utilized for anti-personnel rather than anti-materiel purposes. Such UAVs range in size from small flying insects to that of palm-sized devices equivalent in size to small birds.[86] Such micro-UAVs can be outfitted with a host of traditional weapons—small guns and explosives—as well as unconventional armaments such as incapacitators, poisons, and nerve agents.[87] In this scenario variant, these devices would be intended for urban terrain with the micro-swarms utilized against opposing U.S. soldiers. Of concern is the Anti-Access/Area Denial potentials of such projected micro-drone swarms; however, these UAVs are presently far too sophisticated for insurgent groups to develop, much less field.

It is assumed that over time the dozen or so autonomous drones that can presently self-organize will continue to increase in size into larger and larger groupings. At what point a large enough grouping of drones technically becomes a "swarm" is undefined. From a biological perspective, honeybee swarms range from a few thousand bees into the low tens of thousands. As an arbitrary threshold, real drone swarms might therefore be said to require at least a thousand or so individual UAVs. The operational fielding of such swarms on the battlefield may be viable within 2 decades.[88] While the threat represented by a few hundred weaponized drones is significant, thousands of autonomous self-organizing UAVs operating on the battlefield would have immense U.S. national security implications.

It should be noted, however, that the futures represented by Threat Scenario 3 may or may not be beyond the capability of a terrorist or insurgent group devoid of state or multinational sponsorship. As a result, this capability from a threats perspective may only be available to advanced states such as China or Russia, technologically sophisticated multinational corporation equivalent powers, or their terrorist or insurgent proxies.

MILITARY IMPLICATIONS AND POLICY RESPONSE

Derived from the UAV use threat scenarios, three levels of military significance are foreseen with terrorist and insurgent activities associated with these devices. Because of the technologies that will eventually be associated with UAVs—robotics and expert (and artificial intelligence) systems networked together—their significance is projected to increase over time from the tactical to the operational and then to the strategic levels of concern. These levels of military implication and the suggested policy responses related to them follow.

Tactical.

The impact of even singular terrorist UAV use at this level is viewed as an immediate- and near-term problem. It may represent more of a domestic security issue than an overseas basing or deployment threat—although such weaponized devices could just as easily be utilized for terrorism purposes overseas against service personnel and their families as they could be used against civilians in the United States. The tactical level threat derived from the drone-up shooting, IED crowd targeting, and aircraft takedown scenario variants will be of concern to domestic law enforcement, homeland security, and FBI Joint Terrorism Task Force elements.

For the U.S. Army, the tactical implications of such UAV use will fall within force protection, counterterrorism, and defense support of civil authorities' missions. It will focus on UAV detection, countermeasures, and tactical response. Like civilian law enforcement, the Army will be required to generate new capabilities to mitigate this potential threat. Such response may be as simple as utilizing shotguns in a skeet shooting role that substitute a hostile UAV for a clay disk and/or the inclusion of slightly more involved shotgun ammunition modifications such as less-lethal rounds (eg., baton or multiple-impact with wire lines) for use in urban environments. Stringing up wires to stop access into open venues or in flight choke points may prove to be other hasty anti-drone protocols that may need to be considered if hostile UAV use becomes evident in an area of operations. More advanced measures being considered by the Army in this regard include a new Barrett 25 millimeter anti-drone rifle that has been developed, and electronic warfare systems that had initially been created for counter-IED use, while the Marine Corps is developing a vehicle mounted high power laser.[89]

Since such UAV threat potentials represent a common problem for local, state, and federal law enforcement as well as the Army and other services, it would be prudent for such groups and agencies to form joint working groups to address the tactical concerns such terrorist use could pose. Allied military and federal policing bodies, such as those belonging to Canada, should be considered for inclusion in such working groups. Additionally, from an Army perspective, the Combating Terrorism Center at West Point might represent a natural choice to track ongoing terrorist and insurgent UAV interest and use while the various branch bulletins (e.g., *Infantry* and *Military Police*) could promote UAV countermeasures and response thinking.

Operational.

This level of impact is insurgency environment focused and pertains to the use of groups of human controlled and semi-autonomous UAVs. The virtual martyrs and drone squadron scenario variants portrayed the various types of flying IED, weapons platform, and human insurgent fighter combined arms hybrid threats that could be encountered at this level of concern. While conceivably representing a present-day operational risk scenario as the technologies exist for insurgents to utilize UAVs in this way, this is much more likely a near futures issue that could still be some years out on the horizon before nonstate opposing forces even contemplate or attempt such attacks.

Since no terrorism component is readily foreseen but rather force-on-force engagements are being focused upon, this is not viewed as a domestic law enforcement and homeland security concern. Rather, it is an Army and allied services expeditionary concern, bridging the tactical into the operational level of impact. This means that experimentation and red teaming is warranted related to threat forces use of UAVs in insurgency type environments. Small scale exercises in which red team virtual martyr and drone squadron groups (utilizing UAVs containing paintball grenades and simulated small arms firing) are pitted against

Army infantry squads and platoons (also armed with marking rounds). These exercises are advocated for threat baseline and projection determination purposes. Competitions between Army personnel with the technical skills to create such commercial threat system based quads and related drone systems and the development of the red team concepts to utilize them operationally should also be considered to support such an effort.

Given the foreseen impact at the operational level, the Army effort to develop such small scale UAV red team exercises might best be initially coordinated via programs at the U.S. Army Command and General Staff College, Fort Leavenworth, KS, and supported by writings meant to generate awareness, discussion, and debate at the *Military Review* level of professional writing.[90] Given ongoing U.S. Naval Postgraduate School interest in UAV experimentation, a liaison to their programs should also be established. Ultimately, the limited scale red teaming exercises advocated could be held at any number of Army training venues such as the National Training Center at Fort Irwin, CA, or the Muscatatuck Urban Training Center, Butterville, IN.[91]

Strategic.

While the drone swarms of normal and micro-sized UAVs projected in this threat scenario may still be a few decades out, and possibly even beyond the capacity of terrorist and insurgent forces to field on their own without state sponsorship, now is the time to attempt to get ahead of such developments and help shape the future combat environment.[92] At a minimum, we may presently be in an inter-war period, as experienced between World Wars I and II, when the various elements leading to a revolution in military affairs took place with the evolution of the tank and supporting arms that resulted in the mass armor and mechanized formations that fought in World War II. In this instance, similar disparate elements, involving robotics, expert systems, artificial and collective (cloud-like based) intelligence, network communications, and 3D and 4D replicators may be ushering in another revolution in land warfare involving both ground and aerial based unmanned vehicles and systems.

Given the strategic impact potentials of such aerial (and ground) drone swarms on the future conduct of war, research and writing is required to be vested at the U.S. Army War College level of analysis and policy formulation. Papers derivative from this present effort and earlier ones are needed, as well as shorter essays written for *Parameters*, to highlight concerns, debates, and insights related to robot autonomy and Landpower.[93] Considerations need to be made concerning arms control regimes related to such autonomous, intelligent, and lethal robotic systems[94] as well as their integration with human soldiers into future force structures if that Army unit composition is elected to be followed, as presently appears to be the national trajectory. Further, given the Joint Force nature of American warfighting, the debate on such autonomous robotic systems and drone swarms needs to be expanded to the other senior service war colleges as well as equivalent level allied nation defense educational institutions.[95]

ENDNOTES

1. The term UAV (unmanned aerial vehicle) is now being used with the newer term UAS (unmanned aerial system) to describe these devices. For the purposes of this monograph, the term UAV and drone will be used interchangeably. However, drones are typically viewed as being fixed wing unmanned aircraft, while many would contend that UAVs have rotors and fly more like a helicopter.

2. About 200,000 drones are sold each month commercially—at about $720 million in sales in 2014—with the market projected to double in 2015. See Barbara Booth, "Is It Time to Buy Your Kid a Drone for Christmas?" CNBC News, December 22, 2014, available from *www.cnbc.com/id/102280825.*

3. Roger Pigott, "Heathrow Plane in Near Miss with Drone," BBC News, December 7, 2014, available from *www.bbc.com/news/uk-30369701*; and Justin Davenport, "Police Alert over Unmanned Drones 'Being Flown Illegally Around London's Landmarks'," *London Evening Standard,* January 6, 2015, available from *www.standard.co.uk/news/london/drone-flying-london-police-alert-landmarks-illegal-london-eye-tower-bridge-9959511.html.*

4. Caroline Wheeler, "Terror Threat Alert: UK's Nuclear Plants Are at Serious Risk of Terrorist Drone Strikes," *Express,* February 23, 2015, available from *www.express.co.uk/news/uk/559718/Nclear-plants-are-at-risk-from-a-terrorist-strike-by-unmanned-drones.*

5. Jim Acosta and Jeremy Diamond, "U.S. Intel Worker Blamed for White House Drone Crash," CNN News, January 27, 2015, available from *www.cnn.com/2015/01/26/politics/white-house-device-secret-service/*; and Faine Greenwood, "Man Who Crashed Drone on White House Lawn Won't Be Charged," *Slate,* March 18, 2015, available from *www.slate.com/blogs/future_tense/2015/03/18/white_house_lawn_drone_the_man_who_crashed_it_there_won_t_be_charged.html/.*

6. See "Charlie Hebdo Attack: Three Days of Terror," BBC News, January 14, 2015, available from *www.bbc.com/news/world-europe-30708237*; "Mysterious Drones over Paris Cause Panic," *The Telegraph,* February 24, 2015, available from *www.telegraph.*

co.uk/news/worldnews/europe/france/11431330/Mysterious-drones-over-Paris-cause-panic.html; and "Al Jazeera Journalists Arrested in Connection with Drones over Paris," NBC News, February 25, 2015, available from *www.nbcnews.com/news/world/al-jazeera-journalists-arrested-connection-drones-over-paris-n312776.*

7. The initial documents concerning the emerging UAV terrorism threat are from the 2004-05 period: Michael Gips, "A Remote Threat," *Security Management Online*, October 2002; Eugene Miasnikov, *Threat of Terrorism Using Unmanned Aerial Vehicles: Technical Aspects*, Moscow, Russia: Center for Arms Control, Energy and Environmental Studies at MIPT, June 2004, Translated in English—March 2005, available from *www.armscontrol.ru/uav/report.htm*; and Jay Mandelbaum and James Ralston *et al.*, *Terrorist Use of Improvised or Commercially Available Precision-Guided UAVs at Stand-Off Ranges: An Approach for Formulating Mitigation Considerations*, ADA460419, Alexandria, VA: Institute for Defense Analysis, October 2005, available from *oai.dtic.mil/oai/oai?verb=getRecord&metadataPrefix=html&identifier=ADA460419.*

8. Significant contributions in these areas include Manuel Castells, Information Age: Economy, Society and Culture trilogy—*The Rise of the Network Society* (1996), *The Power of Identity* (1997), and *End of Millennium* (1998)—Cambridge, MA: Blackwell Publishers; and the more recent Erik Brynjolfsson and Andrew McAfee, *The Second Machine Age: Work, Progress, and Prosperity in a Time of Brilliant Technologies*, New York: W. W. Norton & Company, 2014.

9. One of the dominant works in this area is from a Brookings Institute scholar. See P. W. Singer, *Wired for War: The Robotics Revolution and Conflict in the 21st Century*, New York: Penguin, 2009. The Center for New American Security has also been involved in a more recent series on robots on the battlefield. See Paul Scharre, *Robotics on the Battlefield - Part I: Range, Persistence and Daring*, Washington, DC: Center for New American Security, May 21, 2014, available from *www.cnas.org/range-persistence-daring#.VRR4yEuIzFI*; Robert O. Work and Shawn Brimley, *20yy: Preparing for War in the Robotic Age*, Washington, DC: Center for New American Security, January 2014, available from *www.cnas.org/sites/default/files/publications-pdf/CNAS_20YY_WorkBrimley.pdf*; and Paul Scharre, *Robotics on the Battlefield Part II: The Coming Swarm*, Washington, DC: Center for New American Securi-

ty, October 2014, available from *www.cnas.org/sites/default/files/publications-pdf/CNAS_TheComingSwarm_Scharre.pdf*.

10. Steven Metz, "Strategic Insights: The Landpower Robot Revolution Is Coming," Strategic Studies Institute, U.S. Army War College (USAWC) website, December 10, 2014, available from *www.strategicstudiesinstitute.army.mil/index.cfm/articles//Landpower-Robot-Revolution/2014/12/10*. Recent Strategic Studies Institute interest in this topic dates to at least the Robotics & Contemporary/Future Warfare panel, 20th Annual Strategy Conference, "Strategic Implications of Emerging Technologies," USAWC, Carlisle, PA, April 14-16, 2009. Earlier USAWC works on this topic include Mark L. Swinson, *Battlefield Robots for Army XXI*, 1997; and Vincent J. Van Joolen, *Artificial Intelligence and Robotics on the Battlefield of 2020?* 2000.

11. P. W. Singer, *Wired for War: The Robotics Revolution and Conflict in the 21st Century*, New York: Penguin Books, 2009, pp. 196-197.

12. Medea Benjamin, *Drone Warfare: Killing by Remote Control*, London, UK: Verso, 2013, p. 13.

13. Paul J. Springer, *Military Robots and Drones: A Reference Handbook*, Santa Barbara, CA: ABC-CLIO, 2013, pp. 11-12.

14. Singer, pp. 54-56.

15. Springer, pp. 177-178, 189-196.

16. *Ibid*., pp. 209-210.

17. Of these UAVs, 5,346 are RQ-11 Ravens utilized by the Army, Navy, and Special Forces. Overall estimates are derived from Dyke Weatherington, *Current and Future Potential for Unmanned Aircraft Systems*, Office of the Under Secretary of Defense for Acquisition, Technology, and Logistics (OUSD[AT&L]), Unmanned Warfare, briefing, December 15, 2010. See Jeremiah Gertler, *U.S. Unmanned Aerial Systems*, R42136, Washington, DC: Congressional Research Service, January 3, 2012, pp. 2, 11-12. For more on these systems, see Office of the Secretary of Defense, *Unmanned Aircraft Systems Roadmap: 2005-2030*, Washington, DC: Department of Defense, 2005.

18. Greg Miller, "CIA Seeks to Expand Drone Fleet, Officials Say," *The Washington Post*, October 18, 2012, available from *www.washingtonpost.com/world/national-security/cia-seeks-to-expand-drone-fleet-officials-say/2012/10/18/01149a8c-1949-11e2-bd10-5ff056538b7c_story.html*.

19. Gertler, pp. 7-9.

20. See Gips; and James M. Smith, "Aum Shinrikyo," Eric A. Croddy, James J. Wirtz, and Jeffrey A. Larsen, eds., *Weapons of Mass Destruction: An Encyclopedia of Worldwide Policy, Technology, and History*, Santa Barbara, CA: ABC Clio, 2005, p. 32.

21. Prior to the March 1995 subway attack, an earlier sarin gas attack occurred in Matsumoto in June 1994 when the agent was placed in the back of a truck—likely this was the actual attack in which helicopters would have been used had they not failed in testing. See Kenneth Pletcher, "Tokyo Subway Attack of 1995," *Encyclopedia Britannica*, November 2, 2014, available from *www.britannica.com/EBchecked/topic/1669544/Tokyo-subway-attack-of-1995*.

22. Gips.

23. Eugene Miasnikov, *Threat of Terrorism Using Unmanned Aerial Vehicles: Technical Aspects*, Moscow, Russia: Center for Arms Control, Energy and Environmental Studies Moscow Institute of Physics and Technology, 2005, p. 26; and Testimony of Dennis M. Gormley, Senior Fellow, Monterey Institute's Center for Nonproliferation Studies, Before the Subcommittee on National Security, Emerging Threats, and International Affairs of the U.S. House of Representatives Committee on Government Reform, March 9, 2004, available from *cns.miis.edu/testimony/testgorm.htm*.

24. Miasnikov. The notation is "In June 2002, quoting a German intelligence official, the Reuters news agency reported that al Qaeda might be planning to attack passenger aircraft using model airplanes." Cited by Gips.

25. "Colombia - FARC Drones Discovered," EFE News Service, August 28, 2002, in Miasnikov, p. 25.

26. *Ibid.* Also see "Arafat's New Terror Weapon: Exploding Toy Planes," *Debka File*, January 14, 2003, available from *www.debka.com/article/2785/Arafat-s-New-Terror-Weapon-Exploding-Toy-Planes.*

27. "Leaders Reschedule Summit for Next Week," *The Washington Times*, March 10, 2014, available from *www.washingtontimes.com/news/2004/mar/10/20040310-101149-4514r/?page=all.* The specific passage reads:

> Israel's Shin Bet security service said it has arrested a Hezbollah operative who was planning to use a remote-controlled airplane packed with explosives to attack a Jewish settlement. Shadi Abu Alhazin, 22, a resident of the Khan Younis refugee camp in Gaza, began building the plane in 2002, the agency said in a statement. He was arrested in December.

28. Associated Press, "Hezbollah Says It Has Capability to Bomb Israel from the Air," *Haaretz*, November 12, 2004, available from *www.haaretz.com/news/hezbollah-says-it-has-capability-to-bomb-israel-from-the-air-1.139975.*

29. Reuters, "Hezbollah Flies Drone over Northern Israel," *ABC News*, April 12, 2005, available from *www.abc.net.au/cgi-bin/common/printfriendly.pl?http://www.abc.net.au/news/newsitems/200504/s1343190.htm.*

30. Riaz Khan, "Pakistan Army Destroys al-Qaida Hide-Out," *Free Republic*, September 13, 2005, available from *www.freerepublic.com/focus/f-news/1483677/posts.*

31. United States v. Ali Asad Chandia, also known as Abu Qatada and Mohammed Ajmal Khan, also known as Abu Khalid. Indictment. Criminal No. 1:05CR401, The United States District Court of Virginia, Alexandria Division. Filed September 14, 2005.

32. David Eshel, "Israel Intercept Two Attack UAV Launched by Hezbollah," *Defense Update*, August 7, 2006, available from *defense-update.com/2006/08/israel-intercept-two-attack-uav.html.*

33. "Ohio Man Pleads Guilty to Conspiracy to Bomb Targets in Europe and the United States," Washington, DC: U.S. Department of Justice, June 3, 2008, available from *www.justice.gov/archive/opa/pr/2008/June/08-nsd-492.html*; and Paul Cruickshank and Tim Lister, "Analysis: Model Planes as Weapons of Terror," CNN News, September 29, 2011, available from *www.cnn.com/2011/09/29/opinion/model-plane-attack/*. See also the 3:19 minute video embedded in the article.

34. Cruickshank and Lister. Also see Denise Lavoie, "Rezwan Ferdaus Admits Guilt in Plot to Blow Up Pentagon and U.S. Capitol," *Huffington Post*, July 20, 2012, available from *www.huffingtonpost.com/2012/07/20/rezwan-ferdaus-capitol-pentagon_n_1690755.html*.

35. David Cenciotti, "Photo: Is This the First Taliban-Made Drone, Ever?" *The Aviationist*, May 19, 2012, available from *theaviationist.com/2012/05/19/taliban-drone/*.

36. Yaakov Lappin, "IAF Shoots Down UAV in Northern Negev," *The Jerusalem Post*, November 6, 2012, available from *www.jpost.com/landedpages/printarticle.aspx?id=286845*; and "Hezbollah Drone Photographed Secret IDF Bases," *The Jerusalem Post*, October 14, 2012, available from *www.jpost.com/Defense/Hezbollah-drone-photographed-secret-IDF-bases*.

37. "IAF Intercepts UAV in Israeli Airspace," *IDF Website*, April 25, 2013, available from *www.idf.il/1283-18842-EN/Dover.aspx*.

38. Avi Issacharoff, "PA Forces Thwart Hamas Attack Drone Plot in West Bank," *The Times of Israel*, October 25, 2013, available from *www.timesofisrael.com/pa-forces-uncover-hamas-attack-drone-plot/*.

39. Peter Enav, "Hamas Boasts New Level of Sophistication, Releasing Video Showing One of Its Drones for First Time," *National Post*, July 14, 2014, available from *news.nationalpost.com/news/israel-says-it-shot-down-hamas-launched-drone-four-palestinians-killed-in-separate-airstrike*; and Gili Cohen, "Hamas Has More Drones Up Its Sleeve, Defense Officials Say," *Haaretz*, July 15, 2014, available from *www.haaretz.com/news/diplomacy-defense/.premium-1.605140*.

40. Peter Bergen and Emily Schneider, "Now ISIS Has Drones?" CNN News, August 25, 2014, available from *www.cnn.com/2014/08/24/opinion/bergen-schneider-drones-isis/*; and Yasmin Tadjdeh, "Islamic State Militants in Syria Now Have Drone Capabilities," *National Defense*, August 28, 2014, available from *www.nationaldefensemagazine.org/blog/Lists/Posts/Post.aspx?ID=1586*.

41. Scott Shane and Ben Hubbard, "ISIS Displaying a Deft Command of Varied Media," *International New York Times*, August 30, 2014, available from *www.nytimes.com/2014/08/31/world/middleeast/isis-displaying-a-deft-command-of-varied-media.html?_r=0*. See IS drone footage—0.55 minute embedded video.

42. John Hall, "ISIS Propaganda, Call of Duty-Style: Latest Footage Shows Drone's View of Battle-Ravaged Streets of Kobane before Swooping in to Show Gun Battles on the Ground," *Daily Mail*, December 12, 2014, available from *www.dailymail.co.uk/news/article-2871389/ISIS-propaganda-Call-Duty-style-Latest-footage-shows-drone-s-view-battle-ravaged-streets-Kobane-swooping-gun-battles-ground.html*.

43. Adiv Sterman, "Hezbollah Drones Wreak Havoc on Syrian Rebel Bases," *The Times of Israel*, September 21, 2014, available from *www.timesofisrael.com/hezbollah-drones-wreak-havoc-on-syrian-rebel-bases/*; and Peter Bergen and Emily Schneider, "Hezbollah Armed Drone? Militants' New Weapon," CNN News, September 22, 2014, available from *www.cnn.com/2014/09/22/opinion/bergen-schneider-armed-drone-hezbollah/index.html?sr*.

44. In December 2014, an army ground forces commander in Iran, the supplier of Hizbollah UAVs, said that a "suicide drone" capability was now being promoted and, for the first time ever, saw its use in an exercise to attack aerial and ground targets taking place. See "Iran Presents Its Suicide Drones," Special Dispatch 6019, *MEMRI*, April 10, 2015, available from *www.memri.org/report/en/0/0/0/0/0/0/8515.htm*.

45. David Alexander, "U.S. Has Flown 2,320 Strikes against Islamic State at a Cost of $1.83 Billion: Official," Reuters, March 19, 2015, available from *www.reuters.com/article/2015/03/19/us-mideast-crisis-usa-idUSKBN0MF2HC20150319*.

46. This listing may represent only a sampling of the actual incidents that have taken place. According to March 2004 Congressional testimony, "One recent accounting of terrorist activity notes 43 recorded cases involving 14 terrorist groups in which remote-controlled delivery systems were 'either threatened, developed or actually utilized'," Testimony of Dennis M. Gormley, Senior Fellow, Monterey Institute's Center for Nonproliferation Studies, Before the Subcommittee on National Security, Emerging Threats, and International Affairs of the U.S. House of Representatives Committee on Government Reform, March 9, 2004, available from *cns.miis.edu/testimony/testgorm.htm.*

47. Where a potential real time surveillance capability has also been noted is with the domestic U.S. Occupy Movement "Occucopter" which has been utilized during protests since about December 2011. See *hackerspaces.org/wiki/OccuCopter.* Still, the dominant use of both the IS drones and the Occupy drone in these examples appears to be for propaganda purposes.

48. On the hooligan (criminal) side, an example of using a UAV as a form of protest took place at a European 2016 qualifier soccer match in October 2014 between Albania and Serbia. In that match, a drone with a "Greater Albania" flag flew onto the field, which resulted in a riot breaking out. See Tony Manfred, "Albania-Serbia Soccer Brawl Is Unlike Anything You've Seen at a Sporting Event," *Business Insider*, October 15, 2014, available from *www.businessinsider.com/albania-serbia-brawl-2014-10#ixzz3MYdxH5Sh.*

49. Not only do we have a discrepancy in the videos of the attack itself but also in whether this was an explosive tipped drone or a drone mounted standoff weapon. One source of imagery from this incident can be found at "Watch: Hezbollah uses drones against Syrian rebels," *The Jerusalem Post*, September 21, 2014, available from *www.jpost.com/Middle-East/Watch-Hezbollah-uses-drones-against-Syrian-rebels-375986.* This suggests an air-to-ground weapon. A second video attributed to this attack on *YouTube* and *Liveleak* is longer, and shows an attack against individuals rather than a facility. See "Hizbullah Uses (Ayoub) Drone in Arsal (Syrian-lebanon) Border to Attack Al-Nusra Front," *Live Leak*, November 1, 2014, available from *www.liveleak.com/view?i=50f_1414833936.* While a standoff weapon may be plausible, it would be far simpler to utilize an explosive tipped drone.

This new suicide drone capability has been proclaimed by Iran, from whom Hezbollah is acquiring its drones. See Associated Press, "Iran's Army Tests Suicide Drone in Drills," *Military Times*, December 27, 2014, available from *www.militarytimes.com/story/military/2014/12/27/iran-suicide-drone/20938863/*.

50. Milton Hoenig, "Hezbollah and the Use of Drones as a Weapon of Terrorism," *Public Interest Reports*, Vol. 67, Spring 2014, posted June 5, 2014, available from *fas.org/pir-pubs/hezbollah-use-drones-weapon-terrorism/*. A still developing story at the time of the writing of this monograph for publication was a UAV with a radioactive substance on it (cesium) found parked on the roof above the Japanese prime minister's office. See Jake Adelstein, "Drone with Radioactive Material Found on Japanese Prime Minister's Office Roof," *The Los Angeles Times*, April 22, 2015, available from *www.latimes.com/world/asia/la-fg-japan-drone-20150422-story.html*.

51. This Iranian drone image can be found at "Experts on New Iranian Drone: It Can Fly, But that Is It," *The Algemeiner*, May 12, 2013, available from *www.algemeiner.com/2013/05/12/experts-on-new-iranian-drone-it-can-fly-but-that-is-it/*. Basic forensic analysis from the photo by an associate—stand-off weapons expert David Kuhn— provided on November 21, 2014, via email, is as follows:

> This may be a prototype weapon. There are a couple of possibilities. It could be a wire command guided rocket used for low level attack—hence the robust launch canister. We have used tube-launched, optically tracked, wire-guided (TOW) missiles aboard aircraft. This might be a logical start if they are working on a future guided weapon. If that is a launch tube for that or any other purpose, it would be opened ended at the rear. This tube would also keep the rear fins folded until the rocket/missile is fired. The warhead appears to be fused, and part of the fire control system appears to be mounted up on the wing; possibly a sensor. I suspect that their ultimate goal is to create a pylon mounted weapon that has a punch approaching a Hellfire missile. It looks like this drone has a modified 'off the shelf' 360 degree radar under the fuselage, down-looking cameras on the wings, and landing gear that is designed for rough field landings.

52. The video, which shows the successful use of a drone as a platform for a pistol, can be found at Annalee Newitz, "This Video of a Drone with a Gun Will Freak You the Hell Out," *io9*, June 14, 2014, available from *io9.com/this-video-of-a-drone-with-a-gun-will-freak-you-the-hel-513442074*. The original *YouTube* posting was on June 13, 2013. See ClearPlexCorp, "Drone vs Phone: Samsung Galaxy S IV - Drone Strike," *YouTube*, available from *www.youtube.com/watch?x-yt-ts=1421914688&x-yt-cl=84503534&v=jxThXvuP4Vo*. A paintball system mounted on a drone that shoots multiple targets in quick succession also exists. It could conceivably be said to mimic the mounting of a sub-machine gun (utilizing pistol rounds) on a drone. This video can be found at Milo Danger (pseudonym), "Citizen Drone Warfare—Dangerous Information," *YouTube*, December 10, 2012, available from *https://www.youtube.com/watch?v=Jplh7uatr-E*. Note—the operator is using virtual reality (VR) googles for drone flying and weapon targeting purposes.

53. See, for example, Eric Markowitz, "NYPD Scanning the Sky for New Terrorism Threat," CBS News, October 29, 2014, available from *www.cbsnews.com/news/drone-terrorism-threat-is-serious-concern-for-nypd/*. The article includes an embedded video of the newscast related to the topic.

54. A cursory overview shows UAV smuggling incidents taking place at Elmley Prison in Sheerness, Kent, UK (January 2009), a prison in the Tula region south of Moscow, Russia (February 2011), across the Straight of Gibraltar into Spain (Nov 2011), a Brazilian prison (June 2012), various prisons across Quebec, Canada (2013), a prison in Calhoun, Georgia, United States (November 2013), a prison in Melbourne, Australia (March 2014), and in the Kaliningrad region from Lithuania into Russia (May 2014).

55. Robert J. Bunker, "Mexican Cartel Tactical Note #21: Cartel Unmanned Aerial Vehicles (UAVs)," *Small Wars Journal*, August 1, 2014, available from *smallwarsjournal.com/blog/mexican-cartel-tactical-note-21*.

56. Erica Fink, "This Drone Can Steal What's on Your Phone," *CNN Money*, March 20, 2014, available from *money.cnn.com/2014/03/20/technology/security/drone-phone/index.html?section=money_technology&utm_source=feedburner&utm_*

medium=feed&utm_campaign=Feed%3A+rss%2Fmoney_technology+%28Technology%29; and Lauren C. Williams, "New Drone Can Hack into Your Smartphone to Steal Usernames and Passwords," *Think Progress*, March 20, 2014, available from *thinkprogress.org/home/2014/03/20/3416961/drones-hack/*.

57. John Reed, "Marines Get First-Ever Combat Resupply by Drone," *Defensetech*. December 21, 2011, available from *defensetech.org/2011/12/21/marines-get-first-ever-resupply-by-drone/*.

58. Mark Odell, "US Marines Pilot App to Offer Drone Deliveries for Combat Troops," *Financial Times*, May 26, 2014, available from *www.ft.com/intl/cms/s/0/a4cfe44a-e1e0-11e3-915b-00144feabdc0.html#axzz3MlgIjF8T*.

59. See "Pioneer RQ-2A UAV," *Smithsonian National Air and Space Museum*, no date, available from *airandspace.si.edu/collections/artifact.cfm?id=A20000794000*; and Lawrence Burr, *US Fast Battleships 1938-91: The Iowa Class*, New York: Osprey Publishing, 2010, p. 44.

60. A transitional step between line-of-sight to the drone and wearing eyewear is to use a tablet or smart phone as the controller on which the drone video is fed.

61. Robert J. Bunker, "Virtual Martyrs: Jihadists, Oculus Rift, and IED Drones," *TRENDS Research & Advisory*, Terrorism Futures Series, December 14, 2014, available from *trendsinstitution.org/?p=762*.

62. Herve Pellarin, "Drone Racing Star Wars Style Pod Racing Are Back!" *YouTube*, September 30, 2014, available from *https://www.youtube.com/watch?v=ZwL0t5kPf6E*. Also see Elliot Williams, "Quadrotor Pod Racing," *Hackaday*, October 6, 2014, available from *hackaday.com/2014/10/06/quadrotor-pod-racing/*.

63. One example is the use of a storm drone with an iPhone 4 attached to it for aerial video purposes that was for sale on *eBay*. See *www.ebay.co.uk/itm/STORM-DRONE-RC-QUADROTOR-RTF-WITH-DEVO-7-TX-FPV-AERIAL-FILIMING-/121109659563*. Bidding ended on this item on May 22, 2013.

64. The topic of GPS fencing and its application to drone use was discussed by participants at the 2nd Annual Behavioral Informatics and Technology Studies (BITS) research workshop in relationship to a presentation given by the author on threats related to the use of small unmanned aerial vehicles, at the FBI Behavioral Research and Instruction Unit, Critical Incident Response Group, FBI Academy, Quantico, VA, August 25-28, 2014.

65. See the FLIR ONE attachment for IOS and Android devices at *flir.com/flirone/*. The author was given an overview of this device and handled it at the *LA Drone Expo* that took place on December 13, 2014, in Los Angeles, CA. For the recent case of a police UAV detecting a person via their body heat signature, see Eliot Sefton, "Unmanned Police Drone Is Grounded after Arrest," *The Week*, February 16, 2010, available from *www.theweek.co.uk/politics/16585/unmanned-police-drone-grounded-after-arrest*.

66. Flight software exists so that humans are not required to pilot drones—the only requirement is for the user to know how to utilize the software that comes with these systems. See, for example, TOR Universal Control Software, available from *torrobotics.com/post/services/ground-station-software/*.

67. The Naval Postgraduate School is seeking to conduct a large rival swarm-on-swarm demonstration as a proof of the concept. See Debra Werner, "Drone Swarm: Networks of Small UAVs Offer Big Capabilities," *Defense News*, June 12, 2013, available from *archive.defensenews.com/article/20130612/C4ISR/306120029/Drone-Swarm-Networks-Small-UAVs-Offer-Big-Capabilities*. The creation of such networks will be boosted by U.S. Naval unmanned boat swarming advances. See Brendan McGarry, "Navy Reports Breakthrough in Drone 'Swarming'," *Defense Tech*, October 7, 2014, available from *defensetech.org/2014/10/07/navy-reports-breakthrough-in-drone-swarming/*. Such swarming capability is already rapidly progressing among civilian researchers, with Hungarian academics having created 10 self-organizing drones. See Bill Chappell, "Robot Swarm: A Flock Of Drones That Fly Autonomously," *National Public Radio*, February 26, 2014, available from *www.npr.org/blogs/thetwo-way/2014/02/26/283090909/robot-swarm-a-flock-of-drones-that-fly-autonomously*.

68. "Industrial Revolution. 3D Printing Ushers in a New Age," not dated, Southampton, UK: University of Southampton, available from *www.southampton.ac.uk/promotion/3d_printing_02.shtml*. *YouTube* video of the printed aircraft was posted on September 20, 2011. See "Southampton Engineers Fly SULSA the World's First Printed Aircraft," *YouTube*, available from *https://www.youtube.com/watch?x-yt-ts=1421828030&x-yt-cl=84411374&v=nxAjjKkqAQ*.

69. James O'Toole, "Now This Exists: A 3-D-Printed Drone," *CNN Money*, July 8, 2014, available from *money.cnn.com/2014/07/08/technology/innovation/3d-print-drone/*.

70. Jordan Golson, "A Military-Grade Drone That Can Be Printed Anywhere," *Wired*, September 16, 2014, available from *www.wired.com/2014/09/military-grade-drone-can-printed-anywhere/*; and University of Virginia, "The Razor: UVA's 3D-printed U.A.V.," *YouTube*, 29 August 2014, available from *https://www.youtube.com/watch?x-yt-cl=84411374&x-yt-ts=1421828030&v=FwRD7UBGecg*. The professor leading this project had earlier 3D printed out a jet engine.

71. Andy Greenberg, "Meet The 'Liberator': Test-Firing The World's First Fully 3D-Printed Gun," *Forbes*, May 5, 2013, available from *www.forbes.com/sites/andygreenberg/2013/05/05/meet-the-liberator-test-firing-the-worlds-first-fully-3d-printed-gun/*; and Doug Gross, "Texas Company Makes Metal Gun with 3-D Printer," CNN News, November 8, 2013, available from *www.cnn.com/2013/11/08/tech/innovation/3d-printed-metal-gun/index.html*.

72. See Michael Molitch-Hou, "M.I.A.'s New Video Features 3D Printed Guns, Drones," *3D Printing Industry*, May 21, 2014, available from *3dprintingindustry.com/2014/05/21/m-s-new-video-features-3d-printed-guns-drones/*.

73. For a couple of recent bird strike incidents, see Aaron Brown, "Terror Flight: Flybe Jet Loses Engine on Take-Off (but Pilot Flies to Destination Anyway)," *Express*, November 11, 2014, available from *www.express.co.uk/news/uk/534219/Flybe-Passenger-Plane-Engine-Bird-Strike*; and "Flight from Sacramento Makes Emergency Landing after Striking Birds," Reuters, January 9, 2015, available from *www.reuters.com/article/2015/01/09/usa-flight-*

california-idUSL1N0UO1V520150109. Bird strikes, while serious affairs, typically pose very low threats to aircraft survivability because the possibility of such strikes are engineered into engine designs and the birds being ingested by the engines are composed of nondense materials. See, for example, Mark Atwater, "Engineering the Danger Out of Airplane Bird Strikes," *Engineering.com*, November 13, 2013, available from *www.engineering.com/DesignerEdge/DesignerEdgeArticles/ArticleID/6635/Engineering-the-danger-out-of-airplane-bird-strikes.aspx*.

74. These radio controlled hobbyist aircraft can be purchased for less than $1,000.00. For the F-86 craft (1/9 scale) that Rezwan Ferdaus was going to utilize, see *www.jethangar.com/Aircraft/Sabre/F86.html*.

75. A kinetic energy penetrator rod is used to pierce tank armor. It looks like a large pointed steel projectile or dart. In this instance, such a rod would be secured to a hobbyist drone aircraft that would then be used to deliver it into an aircraft engine by flying into it.

76. Such concerns over drones flying over sports arenas have been made known by domestic policing and governmental agencies. See, for instance, Amanda Vicinanzo, "NYPD Developing Strategy to Counter Terrorist Drone Use," *HSToday.us*, November 3, 2014, available from *www.hstoday.us/briefings/daily-news-analysis/single-article/nypd-developing-strategy-to-counter-terrorist-drone-use/4bf0cd619bc5981119cb390b9894968f.html*; and Sean Lawson, "Drone Fear and the FAA Ban on Model Aircraft Near Stadiums," *Forbes*, December 5, 2014, available from *www.forbes.com/sites/seanlawson/2014/12/05/drone-fear-and-the-faa-ban-on-model-aircraft-near-stadiums/*.

77. Such an attack capability traditionally would be carried out using rocket propelled grenades (RPGs) at very low altitudes and surface to air missiles at much higher altitudes. Anti-material (such as .50 caliber) rifles and heavier armaments could also be utilized. None of these are viewed as offering the precision and kinetic mass that a drone strike against a jet engine guided by a video link can provide.

78. Bunker, "Virtual Martyrs."

79. Uncertainty exists if a drone could be rigged with an EFP due to higher weight requirements. A shaped charged variant could conceivably utilize an RPG round-like device, however, in-flight arming safety features would need to be bypassed.

80. If used for terrorism purposes, the multiple drones in these scenario variants would potentially generate much more "terror" than singular drones in the first two variants of Threat Scenario 1.

81. Robert J. Bunker and Charles "Sid" Heal, eds., *Fifth Dimensional Operations: Space-Time-Cyber Dimensionality in Conflict and War—A Terrorism Research Center Book*, Bloomington, IL: iUniverse, 2014.

82. For example, see David McNally, "Army Focuses on Autonomous System Development," *Army Homepage*, November 10, 2014, available from *www.army.mil/article/137718/Army_focuses_on_autonomous_system_development/*.

83. We may see this variant utilized on the battlefield by nonstate forces first, or it may be simply skipped over by nonstate or state militaries that engage in a disruptive technology capability leap and directly field projected swarm forces.

84. Bill Chappell, "Robot Swarm: A Flock Of Drones That Fly Autonomously," *NPR*, February 26, 2014, available from *www.npr.org/blogs/thetwo-way/2014/02/26/283090909/robot-swarm-a-flock-of-drones-that-fly-autonomously*. See the 4:23 minute video.

85. Thomas Claburn, "Navy Tests Swarming Autonomous Boats," *InformationWeek*, November 7, 2014, available from *www.informationweek.com/government/mobile-and-wireless/navy-tests-swarming-autonomous-boats/d/d-id/1316410*. See the 6:38 minute video. The U.S. Navy's Low-Cost Unmanned Aerial Vehicle Swarming Technology (LOCUST) program with its aerial drone launcher represents another advancement in this area. See Dan Lamothe, "Watch LOCUST, the Navy's Prototype Launcher to Send Drones into the Sky," *The Washington Post*, April 14, 2015, available from *www.washingtonpost.com/news/checkpoint/wp/2015/04/14/watch-locust-the-navys-prototype-launcher-to-send-drones-into-the-sky/*. Includes a 1:20 minute video.

86. For a summary of the communication requirements (four building blocks) to create network capabilities for these devices, see Alpen-Adria, "Applications of networked micro-drones," *Phys.org*, March 4, 2015, available from *phys.org/news/2015-03-applications-networked-micro-drones.html.*

87. The U.S. Air Force is presently working on micro-air vehicles for covert access, reconnaissance, targeting, and launching attacks on opposing troops. This service has a "micro-aviary" laboratory in which these robots are being developed. See Conor Friedersdorf, "Like a Swarm of Lethal Bugs: The Most Terrifying Drone Video Yet," *The Atlantic*, February 19, 2013, available from *www.theatlantic.com/technology/archive/2013/02/like-a-swarm-of-lethal-bugs-the-most-terrifying-drone-video-yet/273270/*. See the 4:26 minute video.

88. This general projection is based on a Moore's law estimate derived from increasing computing power that, in turn, doubles autonomous drone swarm size from a base of 12 from 2015 onward. Even this projection may be conservative. A 50-on-50 drone battle was set for later in 2015 at the U.S. Naval Post Graduate School as part of the wider Consortium for Robotics and Unmanned Systems Education and Research (CRUSER) initiative. The current status of this competition is unknown. See Berenice Baker, "Dogfighting Drones—Swarms of Unmanned Battle-Bots Take to the Skies," *Airforce-technology.com*, July 23, 2013, available from *www.airforce-technology.com/features/featuredogfight-drones-unmanned-battle-bot-swarms/*. For DoD views on swarm Autonomous Capability Levels from 1-10 and time frames, see Office of the Secretary of Defense, *Unmanned Aircraft Systems Roadmap: 2005-2030*, Washington, DC: Department of Defense, p. D-10.

89. Brendan McGarry, "Barrett Unveils Drone-Killing Rifle," KitUp! September 26, 2013, available from *kitup.military.com/2013/09/barrett-unveils-drone-killing.html*; "Army Eyes Electronic Warfare Systems to Take Aim at Aerial Drones," *Inside the Army*, October 10, 2014; and Mark Prigg, "The Jeep that Can Down a Drone: US Navy Reveals Anti-UAV Weapon that Can Fire Lasers from a Moving Vehicle," *Daily Mail*, April 16, 2015, available from *www.dailymail.co.uk/sciencetech/article-3042633/The-jeep-drone-Navy-reveals-anti-UAV-weapon-fire-lasers-moving-vehicle.html#ixzz3Xa8vCsSl.*

90. The Foreign Military Studies Office, Fort Leavenworth, KS, is also tracking this issue. See their "Special Look: Counter UAV" section in *OEWatch*, Vol. 5, No. 4, April 2015, available from *fmso.leavenworth.army.mil/OEWatch/201504/index.html.*

91. Additionally, these exercises could be integrated with the field testing of new technologies to combat WMD carrying UAVs being solicited by the Pentagon. See Patrick Tucker, "The Military Wants New Technologies To Fight Drones," *Defense One*, November 6, 2014, available from *www.defenseone.com/technology/2014/11/military-wants-new-technologies-fight-drones/98387/.*

92. Such drone swarms are now becoming a recognized threat to future Army operations. See Michael Peck, "How the U.S. Army Plans to Defeat the Unthinkable: Drone Swarms," *The National Interest*, January 18, 2015, available from *nationalinterest.org/feature/how-the-us-army-plans-defeat-the-unthinkable-drone-swarms-12057.*

93. Very little has been written in *Parameters* on this topic over the last 15 years. The most significant work is probably Thomas K. Adams, "Future Warfare and the Decline of Human Decisionmaking," *Parameters,* Winter 2001-02, pp. 57-71, available from *strategicstudiesinstitute.army.mil/pubs/parameters/Articles/01winter/adams.htm.* That author's discussion of projected developments in autonomous warfighting systems is derived from information provided in the prophetic U.S. Department of the Army, *STAR 21: Strategic Technologies for the Army of the 21st Century*, Washington, DC: National Academy Press, 1996.

94. International deliberations on the fielding of such lethal robots is ongoing:

> This week [April 2015], a ban on lethal autonomous weapons systems (LAWS) is being debated at the Convention on Certain Conventional Weapons (CCW) in Geneva. The U.S. delegation has been non-committal on such a ban, and U.S. policy currently permits the Department of Defense (DoD) to pursue the development of LAWS in a responsible manner.

See Steven Groves, "U.N. Conference Debating a Ban on Autonomous Weapons: Understanding Key Issues," Issue Brief

#4385 on National Security and Defense, Washington, DC: The Heritage Foundation, April 16, 2015, available from *www.heritage.org/research/reports/2015/04/un-conference-debating-a-ban-on-autonomous-weapons-understanding-key-issues*.

95. A recent example of Canadian and American bilateral cooperation concerning the exploration of the challenges related to robotics and autonomous systems in land warfare is the *KCIS 2015: Robotics and Military Operations* conference held May 11-13, 2015, in Kingston, Canada, available from *www.queensu.ca/kcis/index.html*.

U.S. ARMY WAR COLLEGE

Major General William E. Rapp
Commandant

*****

STRATEGIC STUDIES INSTITUTE
and
U.S. ARMY WAR COLLEGE PRESS

Director
Professor Douglas C. Lovelace, Jr.

Director of Research
Dr. Steven K. Metz

Author
Dr. Robert J. Bunker

Editor for Production
Dr. James G. Pierce

Publications Assistant
Ms. Rita A. Rummel

*****

Composition
Mrs. Jennifer E. Nevil